VISUAL SCIENCE
ELECTRICITY

Alan Cooper

Silver Burdett Company

Editor Daphne Butler
Design Richard Garratt
Picture Research Caroline Mitchell
Production Susan Mead

First published 1983

Macdonald & Co. (Publishers) Ltd.
Maxwell House
Worship Street
London EC2A 2EN

Adapted and Published in
the United States by
Silver Burdett Company,
Morristown, New Jersey

1986 Printing

ISBN 0-382-06715-0 (Lib. Bdg.)
ISBN 0-382-09999-0

The Library of Congress has cataloged the
first printing of this title as follows:

Cooper, Alan.
  Electricity / Alan Cooper. — Morristown, N.J.: Silver Burdett Co., 1983.
  48 p.: col. ill.; 28 cm. — (Visual science)
  Includes bibliographical references and
index.
  Summary: Text and illustrations explain
the principles and uses of electricity.
  ISBN 0-382-06715-0
  1. Electricity — Juvenile literature. [1.
Electricity] I. Title. II. Series.
QC527.2.C66 1983
537—dc19   83-50223

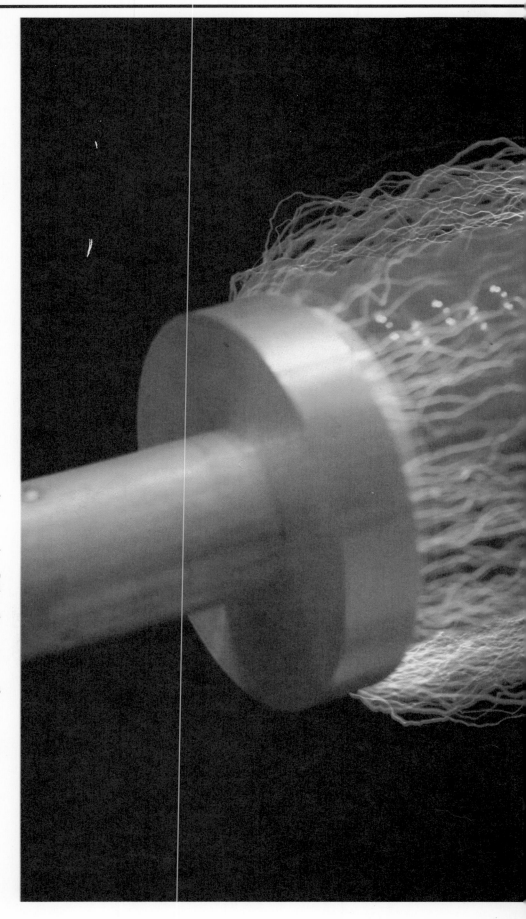

Cover: Las Vegas, U.S.A., by night.
Right: An insulator being tested with
a very high voltage.

Contents

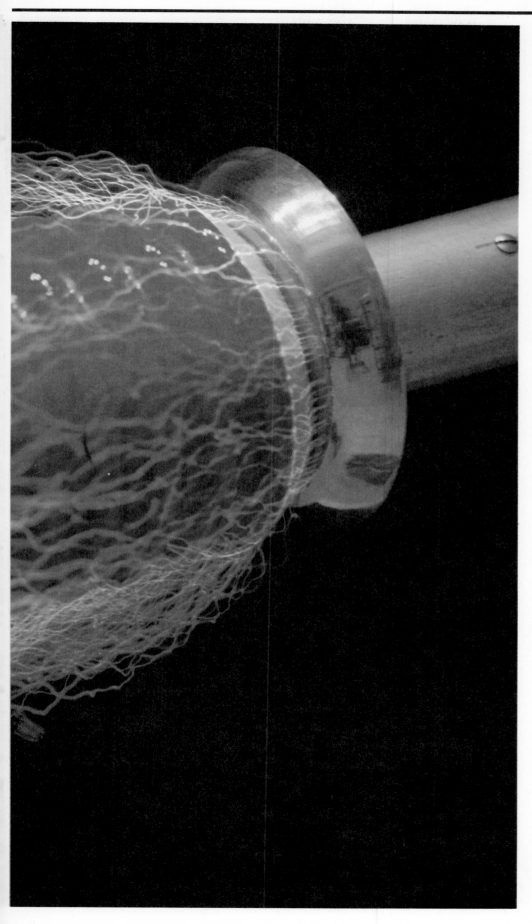

Electric world

Above: This is the Kitt Peak Observatory in Southern Arizona, U.S.A., illuminated by lightning flashes. Each flash only lasts for a fraction of a second but in that time a strong current surges up and down several times between cloud and ground.

Below: The force on the diaphragm of an electrostatic loudspeaker depends on the strength of the electrical voltage which is fed to it from the amplifier. As the voltage varies the diaphragm is pushed backwards and forwards. The air moving through the perforated plates recreates the sound.

Electrostatic speaker

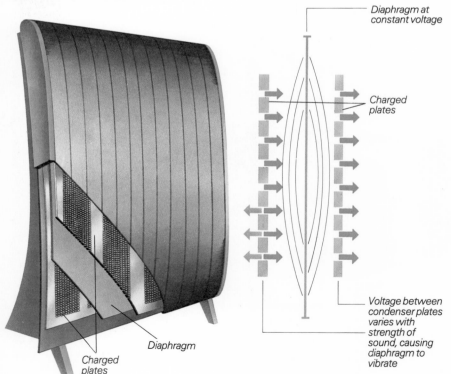

Diaphragm at constant voltage

Charged plates

Voltage between condenser plates varies with strength of sound, causing diaphragm to vibrate

Diaphragm

Charged plates

I counted the number of times I switched on something electrical yesterday. It was thirty-seven, not counting the timers and thermostats which were automatically keeping heaters, refrigerators and blankets at the right temperature. Most of the electrical devices we have all come to rely on have been invented in the last fifty years, and before 1882 there was no public electrical power supply anywhere in the world; not a pylon, not a cable, not a single electric light bulb.

Electricity in nature covers an even wider variety of forms than man's inventions. Lightning which flashes in 12,000 storms a day, the Northern Lights, a dozen varieties of electric fish, magnetic rocks, the uncountable complexity of electrical pulses in the human brain, these are all explained by a few basic principles.

Electrical charge
The factor which is common to all these phenomena, artificial or natural, is electrical charge. The faint crackling when you comb your hair, or take off a nylon or acrylic jersey is due to tiny electrical sparks. The sparks are caused by electrical charge flowing through the air. The air is heated and the atoms torn apart along the path taken by the charge. In the dark you can see the blue glow of the sparks.

Electrical force
Another indication of the charge on the comb is the electrical force which it produces. As long as the charge remains on the comb it will pick up small pieces of paper. That force is not just an amusement for a dry afternoon. It is put to direct use in electrostatic loudspeakers, and in copying machines where light is used to leave charge just within the outlines of the letters on a page. Finely powdered ink sticks to that charge, like paper to a charged comb, and falls off everywhere else. The electrical force is also used in guiding a paint spray on to a car, and ink jets on to paper with such speed and accuracy as to paint in a hundred letters a second. But the importance of the electrical force is immeasurably greater and more fundamental. It is the same electrical force which holds atoms together into groups, called molecules. Every chemical and physical aspect of the matter around us arises from the shapes and strength of molecules. The temperature at which copper boils, the

energy stored in petrol, the sweetness of sugar, and even the processes of life itself, can be related back to the electrical forces between atoms, and between molecules.

Magnetic force

Another natural phenomenon is magnetic force which as we shall see is closely related to electrical force. The cosmic cloud which collapsed to form the Earth and other planets was shaped by magnetic forces, of which a residue acts on magnetic compasses. These are very mild effects. There are stars which show by their radio signals that they are gripped by magnetism hundreds of thousands of times stronger than that of the Earth.

It has taken a thousand years to build up the observations of electric and magnetic effects, which in nature can be quite complicated, to recognise their common factors and reveal that they are all aspects of electricity and magnetism.

In the next few pages we will start with the simplest effects of electrical charge, then see how electrical charge produces electrical current and how currents produce magnetism. Magnetism can, in its turn, produce current.

The list of discoveries and inventions on page 46 shows how quickly electromagnetism was put to use, once it was understood.

Right: The earliest compasses used by sailors were made from a piece of naturally magnetic lodestone floating on a piece of wood. A modern yachtsman uses a more powerful magnet in carefully balanced bearings, but the idea is the same.

Moving charge

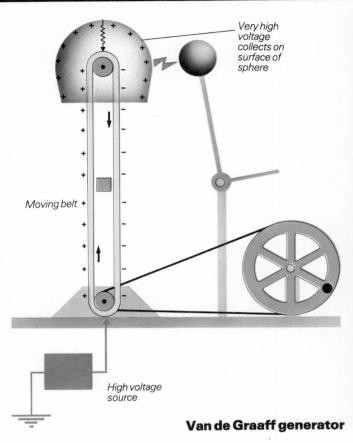

Very high voltage collects on surface of sphere

Moving belt

High voltage source

Van de Graaff generator

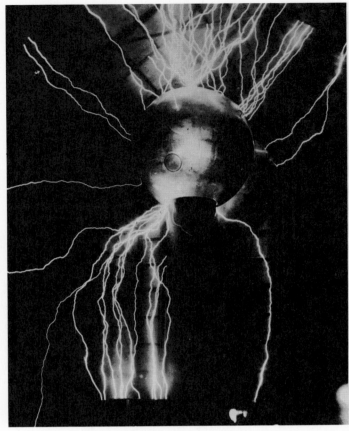

Separating charge

All matter contains enormous amounts of charge of opposite signs, positive and negative. The opposite charges attract each other so strongly that normally there are exactly equal amounts of each, which means zero net charge. When you comb your hair, about a millionth of a millionth of the negative charge in the comb flows out into your hair, and even that is enough to produce sparks.

Even in a machine specially designed to separate out the negative charge, such as the Van de Graaff generator, the fraction transferred is much less than a millionth, yet the effects can be very dramatic.

Moving negative charges make an electric current. In a spark the current is through air (or through gases) from negative to positive and has to make its own path. Metals provide an easy path along which negative charges can flow; therefore a current in a wire is easy to control and can be carried long distances.

Long distance messages

The first use of electricity was in carrying messages; electric light did not come until 40 years later. The Morse code was invented in 1838 by Samuel Morse in the U.S.A. and he put into practice ideas proposed the previous year by Cooke and Wheatstone in England. The Morse Code is still in very common use in radio communication both by amateurs and in ship-to-ship, and ship-to-shore messages.

However, when a way was discovered of converting speech into electricity (by Alexander Bell in 1876), the advantages of being able to send long distance messages became available to all, not only to trained Morse operators. The telephone soon became very popular, and today telephone users in the USA and western Europe can dial directly to 80 per cent of the world's subscribers.

Voltage and resistance

Although both a spark and a current in a wire are flowing charge, they are clearly very different. We must be able to put these differences into words if we are to discuss the subject of electricity further. The easiest way is to take advantage of the similarities between the flow of water and the flow of electricity.

To have a given rate of flow of water in a pipe (say 10 litres per minute) one

Above left: The laboratory version of the Van de Graaff generator shows how an electrostatic charge is built up and discharged. Electrical charge from a high voltage supply is carried upwards by a belt. At the top it is picked off and stored so that the voltage rises steadily until it is so high that a spark carries the charge back to the ground.

Above right: The magnitude of the spark can be seen in the discharge of the large scale version of the generator. The Van de Graaff machine is used to produce millions of volts for particle accelerators.

may either use narrow pipes and a high pressure pump, or wider pipes and a lower pressure pump. In just the same way a given current of electricity can be achieved with a high voltage source (high pressure) through thin wires, or a low voltage source (low pressure) through thick wires. The thin wires have a high resistance to current, and the thick wires have a low resistance. The air has a very high resistance, and sparks can only be produced with very high voltages. All metals have low resistance, but some are lower than others. Silver has the lowest resistance of the common metals, then copper, then aluminium, then iron. Non-metals have very much higher resistance and are never used for carrying

Left: Electricity was used at first not as a power source but to carry messages. The telephone soon became very popular; this is a scene in Chicago only a few years after the telephone was patented by Bell, in 1876.

current long distances. In tiny electrical devices such as transistors, however, the special electrical properties of non-metals are invaluable.

Ohm's Law

The relation between current, voltage and resistance in a circuit can be written down as: voltage equals current multiplied by resistance. This is called Ohm's Law because Georg Simon Ohm first mentioned it in a book he published in 1827. At the time it was ignored but it is now seen to be one of the most important laws in electricity and is worth remembering.

Below: The idea of an electrical circuit can be compared to a water circuit. A battery (pump) provides the voltage (pressure). The current can be turned on or off by a switch (tap). A resistor acts like a coil of narrow bore tubing; if the resistance to the current flow is increased, a bigger voltage (pressure) is needed to maintain the same current.

Electrical circuit

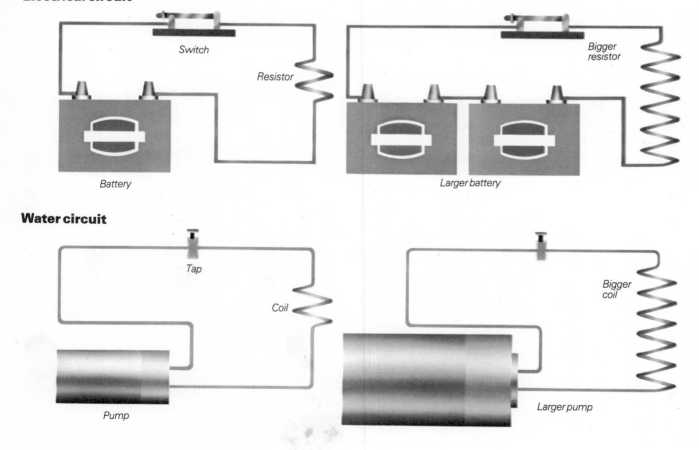

Switch

Resistor

Bigger resistor

Battery

Larger battery

Water circuit

Tap

Coil

Bigger coil

Pump

Larger pump

Electrical Motors

Natural magnets

Nowadays, navigation is an exact science, based on gyrocompasses and networks of radio stations which cover the Earth with a grid of precisely controlled signals. It would be unthinkable to set out across an ocean relying on a 'magic direction finder' bought from a fortune teller. But the first magnetic compasses were made with a naturally magnetic ore (magnetite, originally called lodestone), and the sailors using them cannot have had any idea why or how they worked. There are clear records of the use of magnetic compasses in various parts of the world around 1200 A.D., so the first uses may well have been much earlier.

The magnetism of lodestone and the electric current of lightning flashes are very closely related. But for *600 years* nobody stumbled on the relation between them. Faraday began a systematic, scientific study of electricity and magnetism in 1831, and it took him about 20 years to find the main links between the two.

Magnetic force on a current

It is easy to repeat for oneself the experiments which won Faraday world fame. Magnets can be bought from most hardware stores; possibly because they can be used to pick up dropped pins from behind furniture, but more likely just because they are fun to use and are an integral part of several games.

They come in the form of pairs of steel bars, or in the shape of small steel horseshoes, usually painted red. In either case a piece of iron called a keeper is placed across the ends, and is held in place by the strong attraction of the magnet. The magnetic effect or *field* (Faraday's word) is channelled through the iron while it is in place. But if the iron keeper is removed, the attraction of the magnet for any iron can be felt several centimetres away.

Above: There is no force on a wire between the poles of a magnet until a current flows in the wire. The wire is then deflected by the magnetic force. A coil of wire can be made to rotate, because the force is upwards on one side and downwards on the other. Connecting the current via the split ring makes sure that, in this case, the force is always upwards on the left, and downwards on the right.

Right: Electrical motors are used in a wide range of sizes, from a tiny motor for a clock to those for electric trains such as the Japanese *Bullet Train*.

The motor principle

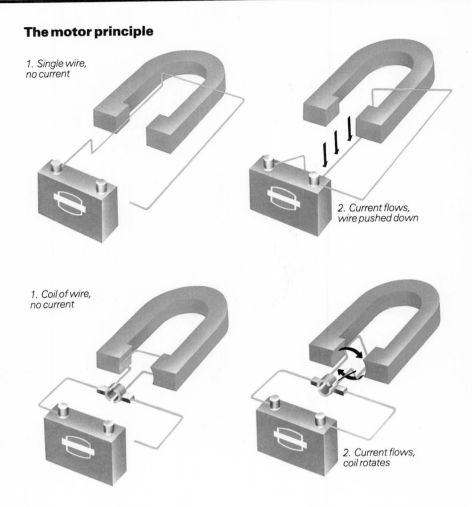

1. Single wire, no current

2. Current flows, wire pushed down

1. Coil of wire, no current

2. Current flows, coil rotates

Such a magnet has no noticeable effect on a piece of copper wire. But if the two ends of the copper wire are attached to a battery, leaving some slack wire so that it can move, it is pushed sideways by the magnet. Since there was no force on the wire itself, the force is on the current running through the wire. But the current is made up of flowing electric charge, so that means that a magnetic field exerts a sideways force on a moving charge. If you have a black and white television set, you can show that the charge which flows to the screen to make the picture can be easily deflected by even a small magnet. But *do not* try this on a colour television set. It spoils the alignment of the colours.

Electric motors

If a wire is bent into a square, so that the current on opposite sides of the square is running in opposite directions, then the forces on the two sides will also be in opposite directions, and so tend to twist the square around its axis. If the wire is taken round a second turn, the force will be twice as big. This is how electric motors are made, with many turns of wire neatly arranged inside a carefully shaped magnet.

Electric motors come in all sizes: from tiny ones to run clocks (or even watches) right up to huge ones to drive locomotives. The efficiency of a motor is the ratio of the mechanical power produced to the electrical power supplied and in most applications should be as large as possible.

Ammeters

The twisting force *due* to a current can be used to *measure* current. Imagine a coil suspended so that it turns against a spring. Then the bigger the current, the bigger the twisting force, and the bigger the angle turned by the coil, all in proportion. An angular scale càn then be marked off in units of current. The unit is called the amp after the French scientist André-Marie Ampère.

Above right: All electrical currents are pushed aside by a magnet. This television is being tested to make sure that the magnetism in the television receiver itself is not high enough to distort the picture.

Right: The turning force of the current in a wire between the poles of a magnet is used in the ammeter to measure that current. The force is measured off against a spring, and read off on the scale.

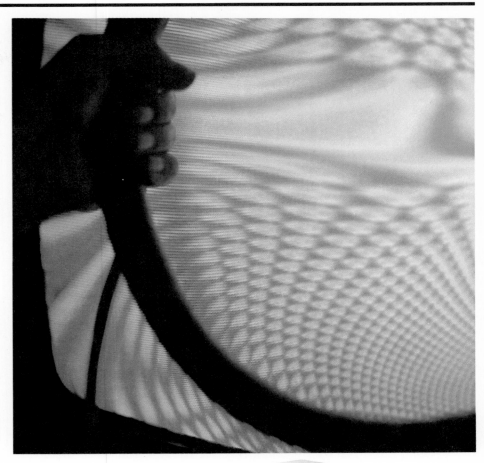

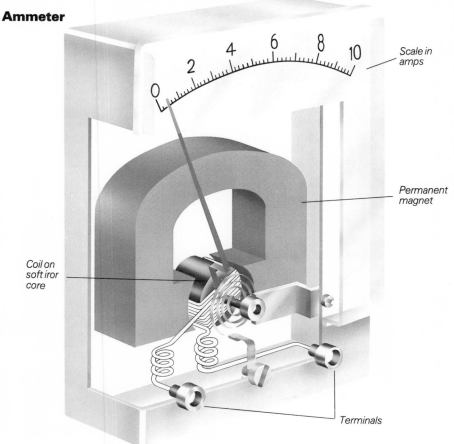

Ammeter

Scale in amps

Permanent magnet

Coil on soft iron core

Terminals

9

Electric magnets

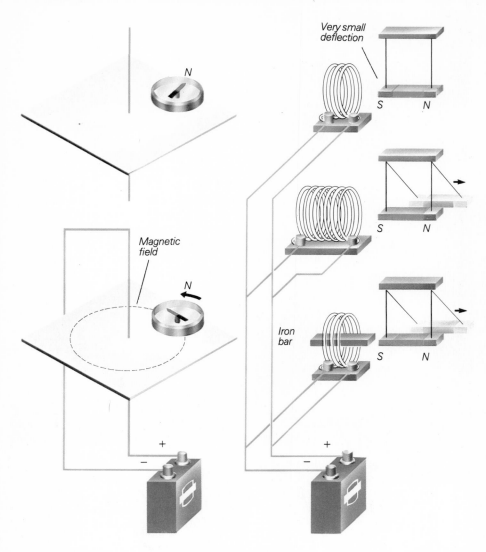

The Earth is a magnet simultaneously attracting all the compasses held by hikers and explorers all over the world. Are there perhaps some vast natural deposits of magnetite which happen to be somewhere near the North and South poles? No, there are no such deposits, and in any case the magnetic poles move considerably (typically 10 km westward per year recently) so this could not possibly be a sensible explanation. The Earth's magnetic field is, in fact, due to electrical currents flowing hundreds of kilometres beneath the surface, in molten rocks which partly thanks to the metals they contain are quite good conductors of electricity. We can estimate how big a current must be flowing inside the Earth by measuring the current needed to deflect a compass only a metre away and multiplying by the distance factor of a few hundred thousand.

Over the years, these currents gradually vary, giving rise to the motion of the magnetic poles. Over longer time scales, the north and south magnetic poles can even swop over. The last time this happened was about 20,000 years ago. Such events have been recorded by man, not intentionally but by chance, arising from the fact that magnetism can be turned on and off by temperature. Below about 400°C magnetism in iron is frozen in, and permanent. Iron at higher temperatures takes up the direction of magnetism of whatever magnetic field it is in; the Earth's field if there is nothing else. In this way, ovens and fireplaces retain magnetism corresponding to the Earth's field when they were last used; perhaps tens of thousands of years ago. A record over a much longer period is provided by rocks welling up from the hot interior of the Earth, and spreading out sideways as they cool.

Magnetic sun

The Sun has a magnetic field, no doubt also produced by internal electric currents. The shape of the Sun's field can be seen at an eclipse, picked out by the

Above: A compass needle points along the Earth's magnetic field. A current flowing in a wire produces a field around the wire which overrides the Earth's field. It makes a pattern of circles round the wire and can be mapped by placing a compass near the wire. Winding the wire into a coil produces a stronger field and a coil can be thought of as a bar magnet. More coils make the magnet stronger. So does putting iron inside the coil.

Magnetic map in sea-bed

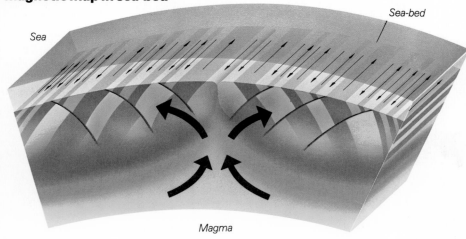

Left: The direction of the Earth's magnetic field reverses every few hundred thousand years, so it has changed half a dozen times during the time man has been in existence. The magnetic effect is frozen into rocks welling up in to the middle of the Atlantic, so keeping a record of the reversals.

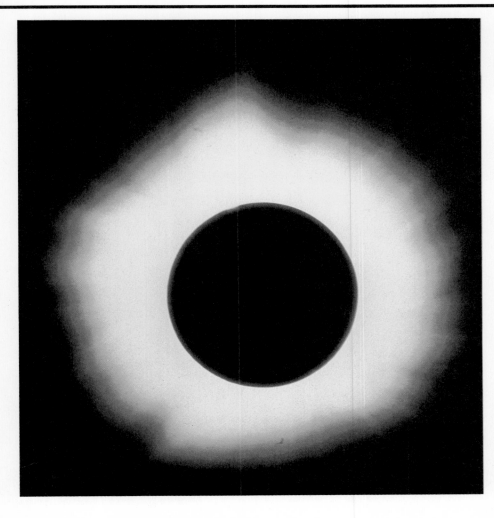

hot gas which streams along following the direction of the field. The Earth's field is a similar shape, but we do not have the benefit of a luminous upper atmosphere to show it up.

Magnetic coils

A turn of wire produces a magnetic field which again has the same shape as that of the Earth or Sun. By making a coil of many turns, each adding its effect to the others, a proportionately stronger field can be produced. A further dramatic increase in field can be achieved simply by putting iron inside the coil. The changes inside iron when magnetised are explained on page 30.

Currents

Two copper wires do not attract each other by themselves or even if a current is flowing in *one* of them. But if a current is flowing in *both* of them, each current is in the magnetic field of the other. If the two currents are flowing in the same direction, the wires are pulled together; if they are flowing in opposite directions the wires are pushed apart. This has to happen because a current produces a magnetic field, and there is a force on a current in a magnetic field. But consider this. Two like charges (two negative charges for instance) repel each other. But move the two charges along parallel to each other and they make two currents travelling in the same direction which attract each other. We might ask how fast do they have to move for the magnetic force pulling the charges together to balance out the electrical force pushing them apart. The first person to answer that question was Maxwell, who set out the laws of electromagnetism in the late nineteenth century. The first person to explain *why* these two forces can cancel each other out was Einstein in 1905.

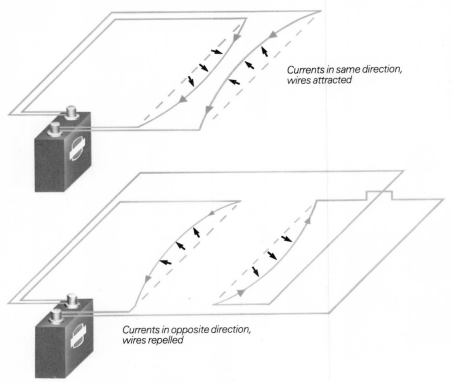

Currents in same direction, wires attracted

Currents in opposite direction, wires repelled

11

Electrical measurement

Voltmeter

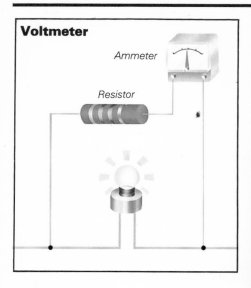

Ammeter

Resistor

Thermocouple

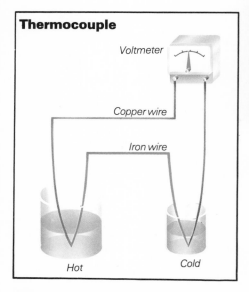

Voltmeter

Copper wire

Iron wire

Hot   Cold

Above left: A voltmeter consists of a resistance and an ammeter connected together. The meter can be marked off in volts, because the current in the resistor connected in parallel with the lamp is proportional to the voltage across the lamp.

Left: A difference in the two temperatures produces a voltage in this circuit and the voltmeter can be used as a thermometer.

Ammeters and voltmeters

There are only two electrical quantities to be measured: current and voltage. Their ratio gives the value of resistance, and their product gives the power used. The electricity meter in every house shows the value of a more complicated quantity: power multiplied by time, which is the total energy supplied. For the measurement of current and voltage, there are two quite different types of meter; those with a needle moving over a scale, and those which show the value directly in figures, like a digital watch. In the window of most electronics shops you will see both. The digital ones will probably be more expensive but they are somewhat easier to use. The ones with a scale are easier to understand, however, because though perhaps compli-

cated to look at they are all based on the simple ammeter described on page 9. Such ammeters can be made very sensitive by using many turns of very fine wire, and a small but powerful magnet. To measure large currents with a sensitive ammeter, most of the current must be bypassed through a resistor in *parallel* with the ammeter. A choice of several resistors allows different ranges of currents to be measured by one instrument.

Voltage is measured by a sensitive ammeter with a high resistance in *series* with it. A multirange voltmeter has several resistors, and a switch with several positions to choose the resistor which gives a convenient position on the scale for the voltage you are measuring. There is a different scale for each resistor.

A multimeter is a versatile instrument which can be switched to measure a large range of currents *and* voltages. Digital multimeters may even have automatic switching of the ranges. Generally multimeters have connections to an internal battery to allow the measurement of resistance as well. You do not have to be an electronic engineer to find it handy to have a multimeter. They are useful in tracking down electrical problems at home. The advantages of electrical measurement go far beyond electrical quantities as such.

Remote measurements

Some measurements need to be made in inaccessible places. If the measurement can be converted to an electrical measurement it can be made very easily because currents pass down thin wires over long distances.

Temperature is a very good example. A motorist needs to keep an eye on the temperature in the car radiator and cannot keep hopping out to stick a thermometer in the tank. Likewise a power engineer cannot climb into a nuclear reactor to make sure it is not overheating. Fortunately there are several ways of changing temperature into a

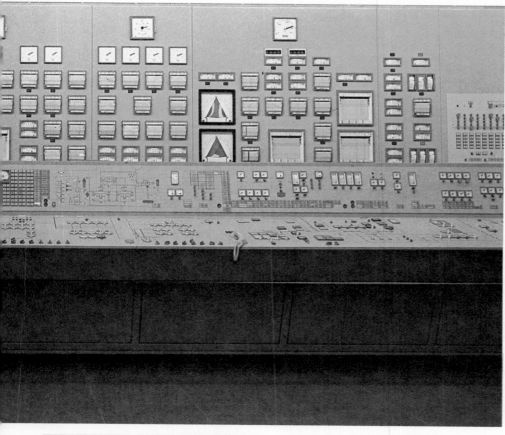

voltage. Two of the most common are the thermocouple and the resistance thermometer.

The level of a liquid, such as petrol in a tank, can be measured by a float. As the float rises or falls it moves across a resistor, and changes the value of its resistance. The current through the resistor therefore depends on the level and can be measured in a convenient place. In the case of a water reservoir the current may be converted to a musical tone so that the water engineer can call the meter by telephone and gauge the level of the reservoir from the pitch of the note. Weather balloons measure temperature, pressure, wind speed and humidity in the atmosphere and relay them to the ground station as a radio signal.

In a large modern factory, key measurements indicating the state of the various stages of a complicated process can be brought together and displayed so clearly that one or two men can precisely control the whole operation. Two extreme examples of remote control come to mind: completely automated machining of engine blocks; and a completely automatic radio telephone exchange on a satellite which is planned in Japan.

Above: A vast number of electrical measurements can be brought together in a control room, such as this one at the Frimmersdorf Power station.

Left: A multimeter is a versatile instrument which can measure current, voltage, or resistance over a wide range according to the settings of the controls. Several designs are available.

Below: When two resistors are connected in series the current is the same in both, so the *voltages* across them are in the same ratio as their *resistances* (25/50 = 50/100). When two resistors are in parallel the current must divide between them in such a way as to give the same voltage across each (2 x 50 = 1 x 100).

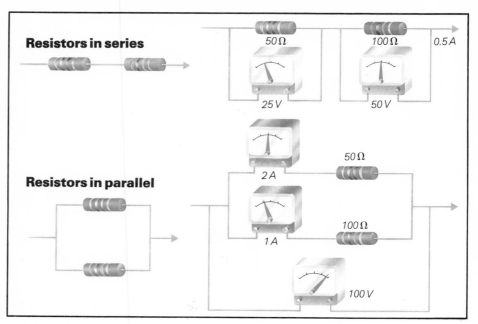

Resistors in series

50 Ω   100 Ω   0.5 A

25 V   50 V

Resistors in parallel

2 A   50 Ω

1 A   100 Ω

100 V

Changing fields

Above: This power station generator is undergoing winding during manufacture.

Below: It does not matter whether a magnet moves in a coil or a coil moves over a magnet, the result is the generation of electricity. This is as true for a laboratory experiment as for a power station generator such as the one shown above.

Power generation

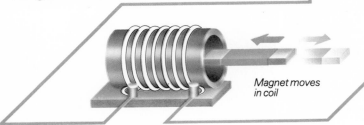

Meter needle moves

Magnet moves in coil

Moving magnets

If you have any sort of ammeter it is easy to show that a moving field produces a voltage simply by connecting a wire directly from the positive to the negative terminal of the meter and moving a magnet (preferably a bar magnet) past the wire. The faster the magnet moves the bigger the kick on the needle. If you have a multimeter you will find that you can produce quite a large current but only a small voltage. The usual trick of making several turns of wire will increase the voltage. Since that also increases the resistance, the current will increase by a smaller factor. This way of generating electricity was discovered by Faraday, and it would be hard to think of any more important or useful scientific result, because it is used in every dynamo, whether in a huge power station or on a bicycle.

The magneto

Small engines for motor cycles, lawn mowers, rotovators, and portable generators need to generate electricity for their spark plugs (and in the case of the motor cycle for lights as well). This is done by a device called a magneto. Magnets are mounted on the rim of a wheel which sweeps them past two or more coils. A crucial point of design is that the coils are wound over iron cores projecting out to a curved pole, which fits very snugly inside the rotating magnets. This exploits the power of iron to guide and strengthen a magnetic field. The effect of the iron is so marked that engineers even talk of magnetic circuits, using the same terms as for current flowing through a wire. The idea is little more than a figure of speech (because nothing flows in a magnetic field) but it helps in the design of motors and dynamos, and some types of pick up used in record players. In the magneto current is induced in the coils and this current is used to power the spark plugs. The wheel holding the magnets is turned by the engine itself once it is running. The energy to start the engine is mechanical; for example you pull a cord or turn a handle.

The microphone

The job of a microphone is to turn the tiny amount of energy of a sound wave into electrical energy. More exactly, the variations in the electrical voltage produced by the microphone must copy

Moving coil microphone

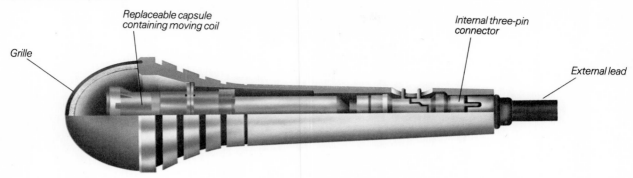

the variation of pressure in the sound wave. The closer the one follows the other, the higher the *fidelity* of the microphone. In the most expensive microphones a diaphragm, vibrated by the sound wave, carries a tiny coil which moves close to a magnet. Moving the coil close to a magnet produces a voltage in the coil. The voltage produced by this sort of microphone may only be a few millionths of a volt. Fortunately a modern transistor amplifier can raise that to any desired level with hardly any loss in fidelity.

The pick-up

A *snapshot* of the sound wave is frozen into the groove of the record and must be recreated as sound as the turntable spins the groove past the pick-up stylus. The stylus, a fine diamond point, rests in the groove, and is vibrated as the record rotates. This seems to be the microphone in reverse, and, indeed,

some pick-ups use the moving coil technique. A more common type of pick-up uses the idea of the magnetic circuit in the following way. The stylus vibrates a magnet. There is a coil nearby, but the voltage is not produced in it directly. The coil has a tiny iron core and as the gap varies between the magnet and the core so the 'resistance' to the 'flow' of the magnetism changes. The magnetic field varies and the voltage is produced in the coil. An amplifier takes the voltage from the coil and produces almost unlimited power for loudspeakers.

Right: A record player pick-up changes the pattern in the grooves on the record to a voltage which can be amplified.

Below: The magneto is a neat generator which provides power for motor cycles and other small machines.

Above: In this microphone sound waves vibrate a small coil inside a magnet. The varying current induced is amplified and can be stored on disc or tape or re-transmitted directly through a loudspeaker.

Fixed magnet pick-up

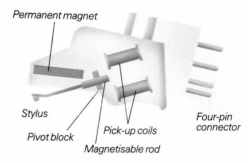

Moving magnet pick-up

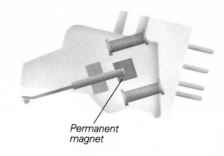

Moving coil pick-up

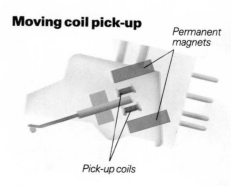

Magneto

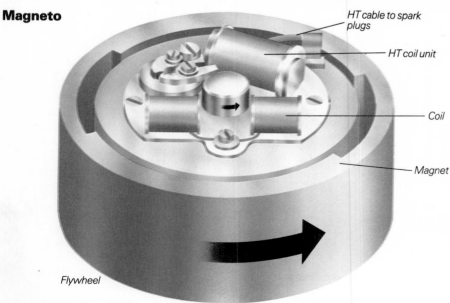

Electrical power

Transformer principle

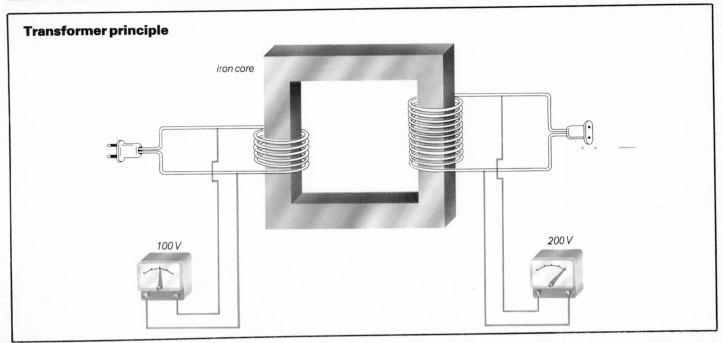

iron core

100 V

200 V

Battery charger principle

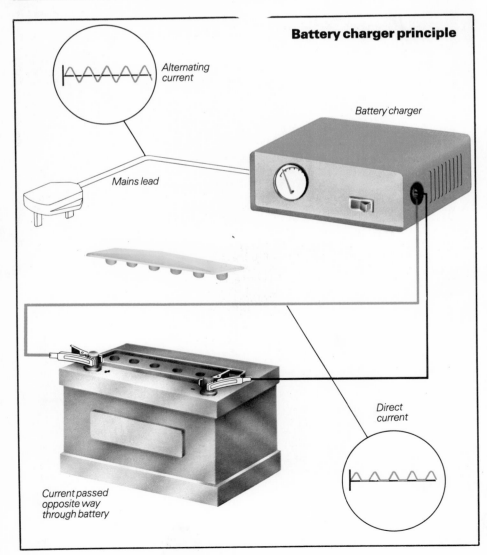

Alternating current

Battery charger

Mains lead

Direct current

Current passed opposite way through battery

Above: One of the main advantages of alternating current is that the voltage of a power supply can easily be changed by using a transformer. The ratio of the incoming voltage to the outgoing voltage is equal to the ratio of the number of turns of wire on the input side to the number of turns on the output side.

Left: A battery charger transforms the mains voltage to a voltage slightly larger than the battery voltage, and then rectifies the current so that it only flows in one direction.

Changing fields

When a magnet moves past a wire which is part of a circuit, a current flows in that circuit. A coil carrying a current produces a magnetic field just like a magnet and can be used to produce a voltage in a wire in the same way as a magnet. But it is not necessary to *move* the coil. If the current in the coil *changes* the magnetic field also changes and a voltage is produced in the wire just the same. Faraday discovered that changing fields produce voltages and his simple experiments were of great *practical* importance because he had in the process constructed the first transformer.

Transformer

A working transformer has an iron circuit to channel the field from one coil through the second one. This is so efficient that there is a simple relation between the voltages across the two coils: they are simply proportional to the number of turns. (Without the iron, magnetic field leaks away, and the voltage produced will be less than the simple rule says).

Being able to control and transform voltages in this way is very useful, but we must remember that it is only a *changing* magnetic field which produces a voltage. While the magnetic field is increasing the current in the pick-up coil is in one direction; while the field is *decreasing* current is produced in the *opposite* direction. It is impossible to keep the field increasing for ever, with the current always in the same direction. Instead, the magnetic field is alternately increased and decreased. In this way there is always current in the second coil, but half the time in one direction and half the time in the other. Such current is called alternating current or just a.c. The current from a battery, which is always in the same direction, is called direct current or d.c.

The advantages of being able to use transformers are so enormous that practically all industrial and domestic power is a.c.

Using a.c.

The alternating voltage produced by all power stations has the smooth variation known as the sine curve, familiar to musicians as a pure tone and to anyone with a calculator as a function of an angle. The sine curve has so many nice features it was the natural choice. The frequency of the a.c. is the number of cycles it makes in a second. Nowadays the unit cycles/second is called the Hertz. Not all countries set up their electricity supply at the same frequency but most countries have adopted either 50 Hertz or 60 Hertz. There was also disagreement on the most convenient voltage for domestic use; Britain chose 240 volts and the U.S.A. 110 volts. The lower voltage is safer but needs thicker wires and twice the current to transmit the same amount of power.

Since voltages can be changed by a transformer, operating an electrical appliance bought in another country using a different voltage is not too difficult. The slightly different frequency however can be a much greater nuisance. Clocks, timers and record player turntables are all geared to work at a particular frequency and usually cannot be changed. However, more and more, such appliances are designed to

generate their *own* frequency with a quartz crystal, but this makes them more expensive. There is also a more subtle problem with a different frequency. The coils and iron cores of transformers are designed for a particular frequency. If operated at the wrong frequency an appliance may run somewhat hotter than expected because it will be less efficient.

Making d.c. from a.c.

Sometimes d.c. is necessary, for charging batteries, and inside radio and television sets for example. This is produced from a.c. with the electrical equivalent of a non-return valve in a water pipe. An electrical one way element is called a rectifier. All it does is cut off alternate peaks, so that the remaining peaks are all in the same direction. To make a steady current a capacitor is used. This stores electricity briefly, and provides current in the times between peaks. The current is not really steady but it is good enough for most purposes.

Below: Transformers are a necessary part of the electricity supply system. High voltages transmitted by overhead cables are reduced for industrial and domestic purposes.

Generating electricity

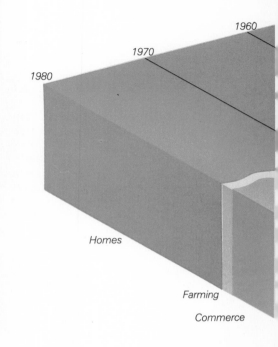

1980   1970   1960

Homes

Farming

Commerce

Industrial power

The rate of technical and industrial advance over the last hundred years, and especially the last fifty years, has been faster than at any other period in man's history. Different countries have been affected sooner or later, to differing extents, and by their own or imported technology, but of the overall effect there can be no doubt. Has it come about because people have suddenly become more intelligent than before, or because government and management is more enlightened? Or because scientific results building up, one upon the other, have suddenly reached levels where they can be useful? Or is it largely for the practical reason that the availability of electrical power, and the new capabilities and facilities it provides, has increased as much as *ten fold* over the last fifty years? This is certainly

a fact in the countries that are currently most industrialised. Which are the causes, and which the effects, are matters for conjecture.

Sources of power

Because electrical power is so easy to transmit over long distances, just about any one of the many sources of energy (see Desmond Boyle's book in this series) can be used for conversion to electrical energy.

Let us imagine a country with 50 million inhabitants and 25 major power stations, so that each power station serves 2 million inhabitants, a major city and extensive surrounding area for instance. If this imaginary country is industrialised and modern, the average power consumption per head will be about 1,000 watts. In this case each power station must provide 2,000

megawatts, corresponding to the largest power stations yet built.

If the country is not yet industrialised, the average consumption is very likely to be much less, more like 100 watts per person. Each power station must then provide only 200 megawatts, which is not big enough to attain the optimum efficiency for a power station.

Coal-powered stations

To provide 2,000 megawatts for a year a coal-powered station uses 600,000 tons of coal. For the 25 power stations, this means about 15 million tons of coal, which will keep many miners busy, not to speak of railwaymen, maintenance engineers for the power lines, engineers for substations, operators of the power station itself, and so on.

For the 200 megawatt station, pro-

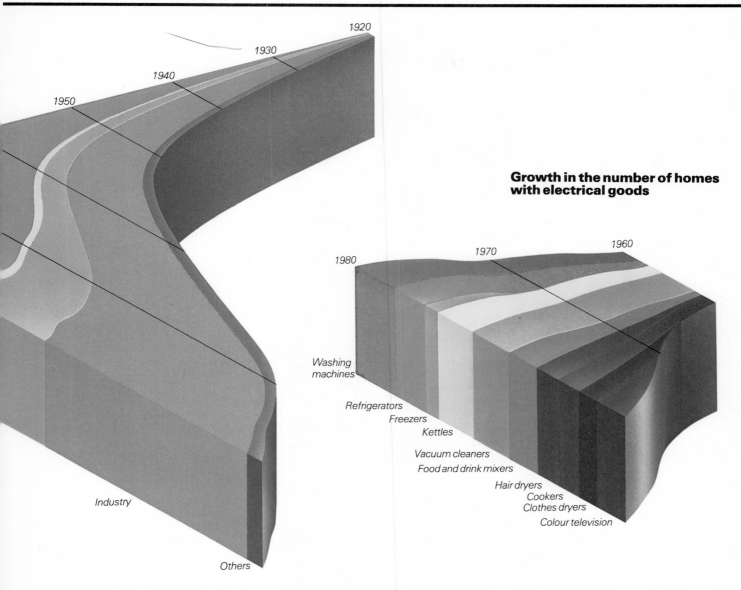

Growth in the number of homes
with electrical goods

1920
1930
1940
1950

Industry

Others

1980
1970
1960

Washing machines
Refrigerators
Freezers
Kettles
Vacuum cleaners
Food and drink mixers
Hair dryers
Cookers
Clothes dryers
Colour television

Above left: The Drax coal-fired power station in the North of England shown with the covers off during testing. The steam comes in at the narrow end and expands through the turbine.

portionately fewer miners and railwaymen will be needed, but the maintenance of the system may be almost as big a job.

Water power
If the country is fortunate enough to have lakes in high mountains, and a good rainfall, it may be able to use a substantial amount of water power. To provide 2,000 megawatts reservoirs about 50 metres deep and covering an area 10,000 square kilometres would be necessary, obviously a major undertaking. Natural lakes are rarely adequate as reservoirs. Some impressive dams have been constructed to

Above: There has been a rapid and continued increase in the use of electricity all through this century. Only in the last decade has the expansion slowed down.

store water as high up as possible, but the best sites have all been used.

There is a specialised use for small reservoirs which solves the problem of storing excess power when demand is low. The problem arises because on the one hand, traditional coal, oil and nuclear powered stations need to run at constant power for maximum efficiency and, on the other hand, modern power sources (tides, wind and waves) naturally have large and uncontrollable variations and rarely match the demand. One solution is to use excess power during low demand to pump water into a small reservoir, and let the water run down again later

in the day to make up when there is peak demand.

Solar power
Although every source of energy can be traced back originally to the Sun, the words 'solar power' usually mean conversion of sunlight into electricity, or using the sun to heat water and the steam to run a generator. There are many substances which do generate a voltage when light falls on them. The photographer's light meter uses such materials, likewise solar panels on satellites. But producing power this way on the same scale as a power station would be very expensive. Alternatively sunlight can be focused on a boiler which runs a steam turbine. The first solar power station of this type was built on a site in Sicily where there is a lot of sun.

The grid

Choosing cables

Here is a problem. 'Big power stations are efficient, but need long cables to take the power where it is needed. Long cables are expensive. The waste heat from thermal stations (coal, oil or nuclear powered) should be used in a 10 mile radius. Choose the most efficient placing of stations and type of cable.' It sounds like a computer game which the power companies would have solved long ago. In fact only a few countries have so far made any attempt to solve the problem of waste heat, which accounts for an unnecessary 50% of the generating cost of electricity, and cable designs are always changing.

The starting point is to find out how high a voltage can be used on the distribution cables without losing too much electricity by discharge into the air. The loss is by a steady flow called corona discharge, which causes the air to glow near to the cable, but stops short of a dramatic flash to the ground. In severe wet weather the discharge is greater. Though rarely bad enough to sheath the wires in a blue glow, it does cause local radio interference. Corona can be avoided by using very thick wires, but that would be far too expensive. Instead the highest voltage lines use a group of three or four wires which gets some of the advantage of a single thick cable at much lower cost. The cables are sufficiently thick for these bunched wires to be seen from the ground.

The advantage of using the highest voltage possible is to keep the current as low as possible for a given power (remember, power equals current multiplied by voltage). Lower currents can be carried on thinner, and therefore cheaper, cables. Of course the cable must not be so thin that it cannot support its own weight. The tension in the wire can be made lower by allowing a greater sag between the pylons, but that could be dangerous in hot weather

Above: Outside cities, power is carried on overhead lines. For the highest voltage, multiple cables and tall pylons are used. Smaller pylons carry electricity out to small towns.

Left: The connections of the grid must be constantly adapted to the needs of the moment. Ordinary switches would quickly be destroyed by sparks; these are oil filled and power driven.

when the cables expand and the sag increases. Sparks to the ground are a troublesome cause of power failures in hot weather. In fact the power carried by a line may have to be reduced in hot weather, which is bad news for those using the power for air conditioners. The only metals with low enough electrical resistance and price are copper and aluminium. But their prices vary and further complicate the problem. The cheapest design this year may be too expensive next year.

The highest voltage lines used in Britian (for the longest links) carry 400,000 volts, and may be recognised by their distinctive pylons and cables. For shorter distances, various voltages are used, 230,000 V, 66,000 V, 33,000 V, and finally 240 V into the home.

Switchyards and substations

The network of lines, linked by transformers down to successively lower voltages is known as the grid. The internal connections of the grid lines must be changed frequently as demand varies across a country during the day, and different power stations close down through failure or the need for maintenance work. There are some impressive substations with massive, and perhaps in its own way rather elegant, switchgear and transformers to perform these adjustments.

Mistakes and failures are bound to occur in such an enormously complicated system. One of the most spectacular occurred in 1965 in the U.S.A. when much of the East Coast lost supplies for a day or more in some parts because a small relay at the Niagara Falls power station had been wrongly programmed. This led to huge amounts of power being swung from one region to another, progressively cutting out safety switches in the substations.

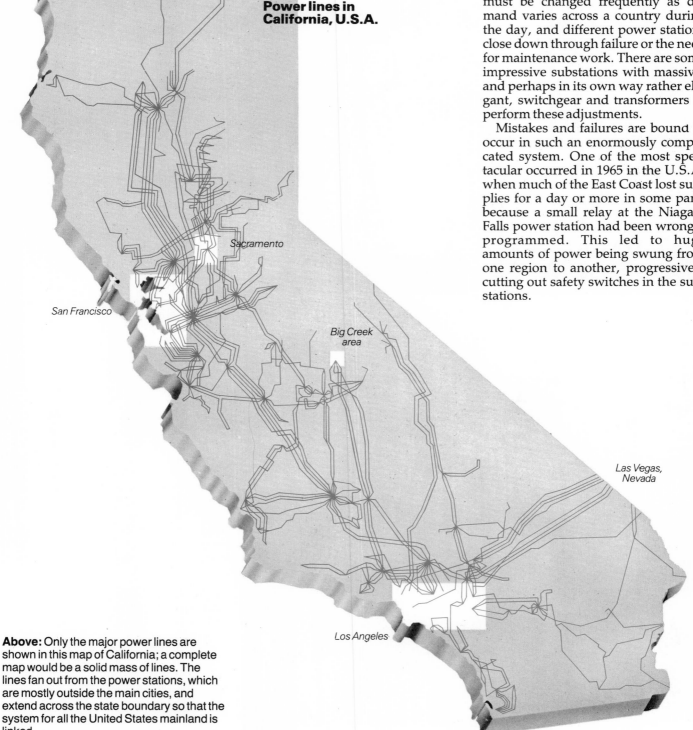

Power lines in California, U.S.A.

Sacramento

San Francisco

Big Creek area

Las Vegas, Nevada

Los Angeles

Above: Only the major power lines are shown in this map of California; a complete map would be a solid mass of lines. The lines fan out from the power stations, which are mostly outside the main cities, and extend across the state boundary so that the system for all the United States mainland is linked.

Circuits and earths

Circuits

Electricity can only flow in a *circuit*: that is a loop of wire which starts from one side of a battery, generator, or transformer, and following an unbroken line, eventually ends up on the other side. This is always true, even though the lines may be long and go under some disguises. Imagine trying to follow a line from the positive side of the battery of a portable radio, through to the negative side. An alternating current circuit for a single lamp seems an easier bet, but what happens to the wire when it has gone through the fuse box? It does *not* go all the way back to the power station, only to the last transformer, usually by underground cable, where it goes round a coil and then back into your house again; an unbroken loop of wire with the transformer coil at one end and the lamp filament at the other end.

Pylon circuits

One would expect overhead lines on pylons to carry pairs of wires, but they turn out to carry three wires, on small pylons or poles, or three thick and one thin wire, or six thick and one thin, on the biggest pylons. To see why look at the diagram on p. 16 of the way alternating current varies. That arrangement is called single phase. Power is distributed on overhead lines in a three phase system, in which the voltage on, and therefore current in, each of three wires varies as shown in the graph. The fourth, thin wire is the return wire. But if it carries the return current from all three other wires, should it not be *thicker* than the others, not thinner? That is the clever part of the system. The sum of the three currents is, in fact, zero. Notice first of all that the three currents are never all in the same direction, then try adding up at any point in the cycle. The three always cancel out. In theory, the return wire is unnecessary. In practice the three coils (or other devices) fed by the three wires will never be *exactly* balanced, so there will be some small current in the return

Right: Current can only flow in a closed circuit. This is clear enough with a battery and lamp. It is also true of a mains supply, but the circuit is completed outside the house at the last transformer. Each line of three-phase supply carried by the grid is a.c. but there is a special relation between them which allows the return wire to be very thin. Try adding up the three currents at any point on the time axis.

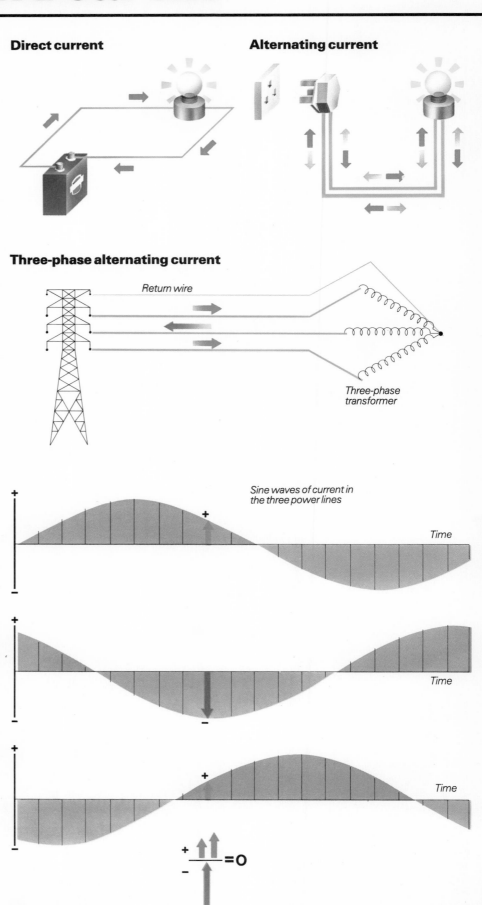

Direct current

Alternating current

Three-phase alternating current

Return wire

Three-phase transformer

Sine waves of current in the three power lines

Time

Time

Time

$$\begin{array}{c} + \\ \hline - \end{array} \uparrow\uparrow \; = 0$$

Right: There is a close analogy between using the Earth itself as a return path for the current, and pumping water out to sea. The electricity finds its way back, as does water, but is spread over a wide area.

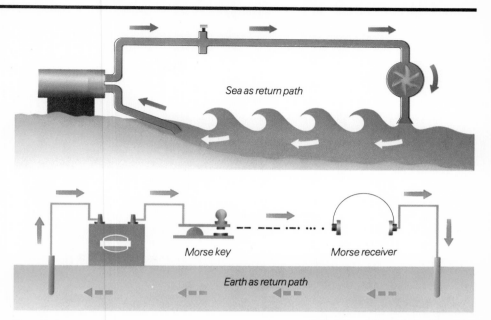

Sea as return path

Morse key   *Morse receiver*

Earth as return path

Earth as a return

RESISTANCE IN OHMS	
Material	**Resistance in ohms 1cm cube**
Copper	0.000 0015
Aluminium	0.000 0025
Iron	0.000 0089
Tungsten	0.000 0049
Carbon	0.004
Paper	1000 000 000 000 (10^{12})
Rubber	1000 000 000 000 000 (10^{15})
PVC plastic	up to 100 000 000 000 000 (10^{14})
Ceramic	10 000 000 000 000 (10^{13})

wire. What about the total power? That had better not be zero otherwise there is no point in having the lines at all. Power is the product of current and voltage (which are always in the same direction). If they are both negative, the product is still positive. The sum of the three powers is thus always positive and, in fact, is constant. This is one of the advantages of three phase a.c. since it means a smooth, constant torque on the motor generators. The thinness of the fourth wire not only saves on the cost of cable but is related to minimising the wear on bearings.

Earth wire

To save on the cost of a return wire, early Morse telegraphy systems sometimes used the earth itself. It may help to think of the analogy of pumping water out to sea and using the sea as a return path. It would be impossible to point out the actual line of the return path through the sea (or earth) but the water (or electrical current) can certainly find its way back. In fact the return electrical current through the earth spreads out over a wide area; that is what keeps the resistance tolerably low. Square millimeter for square millimeter, the resistance of soil or rock is far higher than, say, copper. In the earth there are countless paths in parallel. The analogy with water flow is very close here. Water soaks quickly through porous earth. There are thousands of different pores, that is different paths for the water to take. Each is narrow and tiny, but only carries a little water. The earth is *porous* to electricity in a similar way, and only offers a modest resistance.

Every domestic electrical system includes a plate or pipe sunk into the earth outside the house. The cable to every major appliance has three wires,

one of which (usually coated in insulation coloured green and/or yellow) is connected back to the earth connection. Whilst it might be useful in signalling by Morse to your neighbours, that is not the idea! Instead, the earth wire is connected to the metal casing of the appliance. This prevents you from ever getting a shock from the casing if a fault occurs making it live.

The earth wire

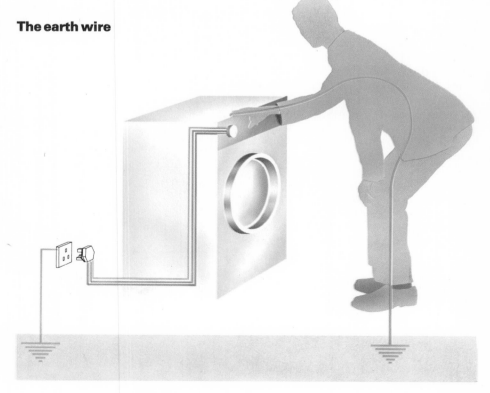

Right: It is standard practice to connect the outside of an appliance to the Earth. Even a faulty appliance cannot then give you an electric shock, provided that your feet are at the same voltage as the Earth.

In the home

Insulation

The earth wire is to protect people from getting a shock from faulty electrical wiring. The casing of an appliance could be 'live' (at a high voltage) if the insulation on the supply wires were broken or worn. Rubber has a very high resistance and was once used for insulation. However it gradually hardens and cracks becoming dangerous. Modern cable has plastic insulation. The difference in resistance between conductors such as copper, and insulators such as plastic, is staggeringly large as can be seen from the table on page 23.

Fuses

A fuse is a short length of thin wire and is used to protect an appliance when it goes wrong, rather than to avoid injury to the person using it. Suppose the incoming live and outgoing neutral wires touch in an appliance, through faulty insulation. Without a fuse an enormous current would flow. There would be overheating, possibly fire, or even an explosion of vaporised metal. In a properly protected circuit it is only the *fuse* which disappears in a puff of smoke. The fuse may be in the plug itself, and always in a fuse box near the entry of the cables into the house. In either case it can be replaced easily, safely and cheaply.

There is a master fuse which carries all the current into the house and cannot be changed without breaking a sealing wire. This fuse belongs to the electricity company and protects the supply wire from the last transformer of the grid. This wire normally has a maximum capacity of about 100 amperes. The householder could misguidedly replace all his worn fuses with such high values that he could draw enough current to melt the incoming wire and perhaps damage the transformer. The company's fuse prevents this. Of course, inserting fuses which do not provide any useful protection makes no sense for the householder either. In the fuse box, each fuse should be rated at a value somewhat higher than the sum of the currents taken by all the appliances likely to be used at the same time on the circuit.

In some cases the protection is provided by a circuit breaker in the place of the fuse. Like a fuse, it breaks the circuit if and when excessive current is drawn, but can be reset simply by pushing a button rather than replacing a wire.

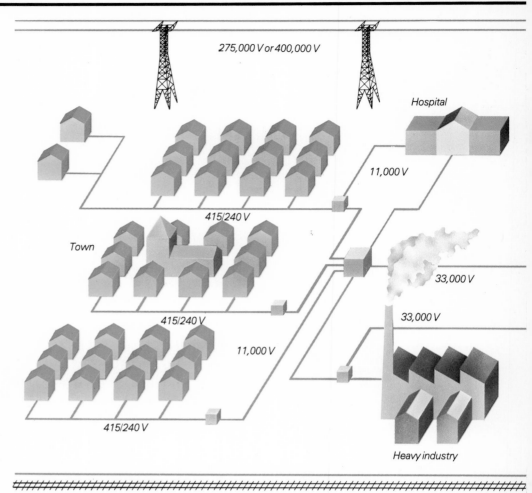

275,000 V or 400,000 V

Hospital

11,000 V

415/240 V

Town

33,000 V

415/240 V

33,000 V

11,000 V

415/240 V

Heavy industry

Above: The power from the main grid is split, through substations, to supply smaller and smaller groups of buildings, and finally to supply your house.

Below: If an appliance is faulty and draws a dangerously high current, a fuse melts and breaks the current. Each circuit in the house must be protected in this way.

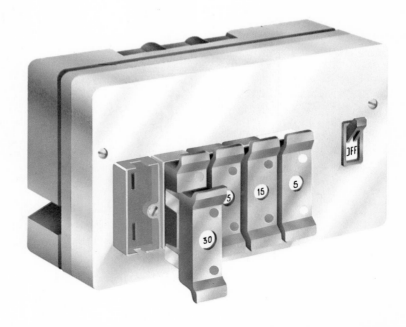

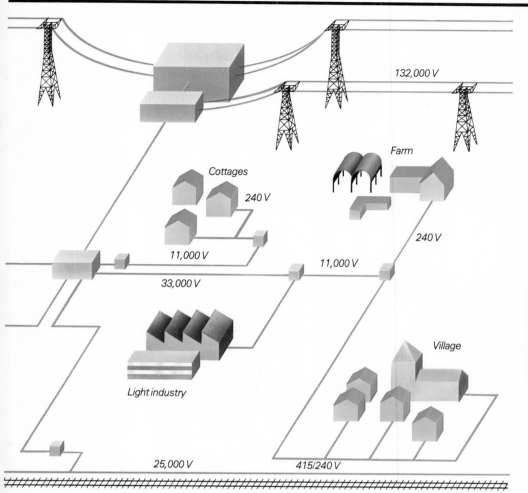

| 132,000 V |
| 240 V |
| 11,000 V |
| 11,000 V |
| 240 V |
| 33,000 V |
| 25,000 V |
| 415/240 V |

Farm
Cottages
Light industry
Village

TYPICAL POWER RATINGS

Appliance	Power needed in watts
Clock	1-5
Portable radio	5-15
Music centre	50-200
Television set	30-300
Refrigerator	100-200
Freezer	200-400
Toaster	1000-2000
Cooker/oven	1500-3000
Washing machine	1500-3000
Dish washer	1500-3000
Vacuum cleaner	200-1000
Lighting for 6 rooms	250-500
Immersion heater	2000-3000

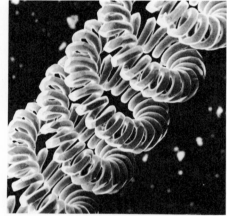

Appliances

Every electrical appliance usually has a tag showing the current and power which it uses. If the current is known the appliance can be protected with the correct fuse. If the power is known the cost of using the appliance can be estimated. If the power is not marked it can be worked out by multiplying the current by the voltage. For instance a toaster taking 5 amps at 110 volts uses 550 watts.

TYPICAL VALUES OF FUSES

Circuit	240 volt supply	110 volt supply
Upstairs lights	5 amps	10 amps
Downstairs lights	5 amps	10 amps
Immersion heater	15 amps	30 amps
Outlets for cooker and washing machine	30 amps	60 amps
Upstairs main outlets	15 amps	30 amps

Right: Incandescent lamps are made in a variety of shapes and sizes some of which are shown here.
Above right: To concentrate the heat the filament is wound into a coiled coil, magnified here two hundred times.

Science and electricity

A pioneering genius

Faraday was born in 1791, the son of a blacksmith, and was fascinated by science from his boyhood days. He became, at the age of 21, an assistant to a respected chemist, Sir Humphrey Davy, and Faraday's first few papers were about some new chemical compounds which he had isolated. From the age of 30, he embarked on a brilliant and original series of experiments which revealed the exact nature of the intimate relations between electricity and magnetism. Later he discovered that he could magnetically influence the polarisation of light. He was not, however, able to work out the relation between electromagnetism and radiation, for the simple reason that he was no good at mathematics and was unable to write down equations to express his own results.

Let there be radio waves

That step was made by Maxwell, born into a prosperous Scottish family and launched into a brilliant academic career by publishing two papers at the age of nineteen, taking a first class degree at Cambridge four years later, and being appointed a professor at the age of twenty-five. This was in 1856, by which time Faraday had completed most of his work and was entering retirement. Maxwell was entrusted with setting up one of the world's most famous physics research laboratories, the Cavendish laboratory in Cam-

bridge, England, but although his achievements were broad based, it was undoubtedly his discovery of the equations of electromagnetism which placed him, with Newton and Einstein, among the great physicists of all time. The equations predicted the existence of electromagnetic waves. Maxwell identified light as one form of such waves and predicted that waves of longer wavelength (radio waves) would be discovered, and be found to travel at exactly the same speed as that of light, a fundamental constant of nature. You can see from the list of dates on p. 46 that although he knew that his theory gave the correct speed for light, he did not live to see his prediction of radio waves verified. It was not until 1888 that Hertz, a German physicist, demonstrated their existence, and it was not until 1894 that a twenty-year-old Italian called Marconi succeeded in sending the first radio signals.

The inventors

Many inventors found fame and fortune by exploiting the deep understanding of electromagnetism achieved by the British physicists Maxwell and Faraday. Marconi, Edison and Bell in the United States were especially productive. The telephone, gramophone, and long wave and short wave radio, all indispensable to modern life, appeared during the last few years of the nineteenth century. An even more

important discovery was made during this time, back in Cambridge, England.

The particles of electricity

In 1897 Professor J.J. Thomson first conveyed a current through a vacuum in a simple piece of apparatus from which the television tube has been developed. He showed that the current was made up of tiny individual particles, electrons, and measured their mass and charge. Soon afterwards the proton was discovered, and measured, using a variant of Thomson's apparatus. The negative charge of the electron and the positive charge of the proton hold them together to form a hydrogen atom. But in metallic conductors, some of the electrons can wander freely, making up an electrical current. In a vacuum tube, the current of electrons can be controlled so precisely as to make powerful amplifiers, working over the complete range of radio and audio frequencies. With these discoveries a deeper understanding of electricity and an unforeseeable panoply of electronic inventions became possible.

Below: Different types of radiation are detected in very different ways, but all are electromagnetic waves differing in wavelength. The only radiation which we can perceive directly is visible light and heat, though the effects of gamma radiation soon become evident on living creatures.

Electromagnetic spectrum

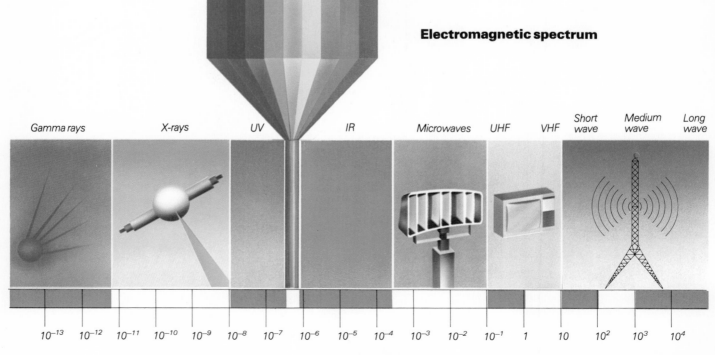

Gamma rays   X-rays   UV   IR   Microwaves   UHF   VHF   Short wave   Medium wave   Long wave

10^{-13} 10^{-12} 10^{-11} 10^{-10} 10^{-9} 10^{-8} 10^{-7} 10^{-6} 10^{-5} 10^{-4} 10^{-3} 10^{-2} 10^{-1} 1 10 10^{2} 10^{3} 10^{4}

(Wavelengths in metres)

Below: All the ideas discussed so far were linked together mathematically by Maxwell. Here some of the links are shown pictorially. Start on the left hand side, with the fact that current is moving charge. Moving a conductor in a field produced a current, and, on the other hand, a magnetic field pushes on a current. Putting this together with the fact that a current produces a magnetic field, Maxwell *predicted* that there must be electromagnetic waves. Such is the power of mathematics.

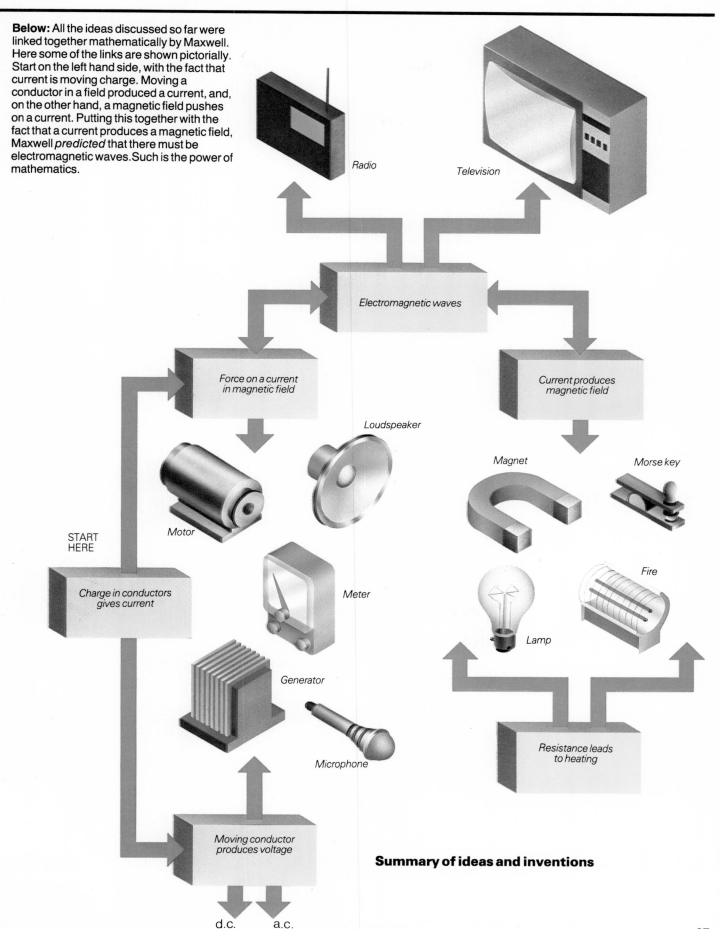

Radio

Television

Electromagnetic waves

Force on a current in magnetic field

Current produces magnetic field

Loudspeaker

Motor

Magnet

Morse key

START HERE

Charge in conductors gives current

Meter

Lamp

Fire

Generator

Microphone

Resistance leads to heating

Moving conductor produces voltage

Summary of ideas and inventions

d.c.   a.c.

27

The working atom

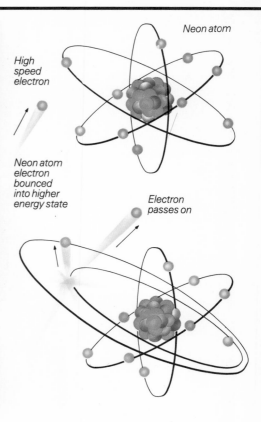

Neon atom

High speed electron

Neon atom electron bounced into higher energy state

Electron passes on

Electron falls back

Energy given off as red light

Atoms

It is the electrical force which holds electrons in atoms, and groups atoms into molecules. The nucleus is thousands of times more massive than the electrons, but is crushed into a tiny speck inside the region of the orbiting electrons by its own, separate nuclear forces. In the same way that it takes more rocket energy to send a satellite to a high orbit than to a low one so the outermost electrons have more energy than the inner electrons. The electrons keep to separate allowed regions, rather as aeroplanes avoid collisions by keeping to distinct air lanes. But there is no analogy for the way an electron can change instantaneously from one 'air lane' to another by giving out a flash of light and dropping to a lower energy region, or by absorbing light and rising to a higher energy region. Such energy changes only occur inside the atom, and consequently we can never see them happen.

Light from atoms

There can nevertheless be no doubt that these energy rules apply inside the atom, because atoms only emit light of special colours, characteristic of the particular atom. Given light of a particular colour we can therefore look up which type of atom it must have come

Above: The red of the neon light dominates Hong Kong at night. Other colours are produced by using different gases in the discharge tubes.

Right: The violent life of a neon atom in the gas inside a fluorescent light. Passing electrons disturb the outer layers of the atom. A burst of light is emitted as the atom readjusts. After battling its way through the tube, an electron slips round the external circuit and comes back for another go.

from. Not all light, however, comes from *inside* the atom. Incandescent bulbs give out a white light which comes from the random jostling and collisions of the atoms in the hot metallic filament.

Converting electricity to light

Let us compare the two types of lamp in common use: incandescent lights, traditionally bulb-shaped, and fluorescent lights, traditionally long tubes. Incandescent lights use a current of electrons passing through a wire which is called a filament. The electrons collide with the atoms of the filament and the energy lost in the collisions appears as heat and light. What should the filament be made of? Early lamps used

carbon, which has a higher resistance than the metals (see page 22), and that means more heating. The hotter the wire the brighter it glows. Good conductors, such as the metals, are those materials whose ordered structure allows electrons to glide through with fewer collisions. At first sight this seems a disadvantage for a lamp filament, but if the wire is made very thin the resistance can be made higher. Imagine the electrons as a crowd trying to get through a narrow exit. The narrower the exit the more jostling there is. Nowadays lamps use tungsten filaments which have to be very thin, but because tungsten is so much stronger than carbon tungsten lamps last much longer.

Fluorescent tubes

Fluorescent lights do not have the high temperature of incandescent lights. They use the light from within atoms and because there is no need to have the atoms jostling around there is little heat. Electric current is passed through a gas in a tube. Electrons attached to the atoms of the gas are knocked into a higher energy orbit by the electrons of the current. An electron never stays in the higher energy state very long. After a tiny fraction of a second it gives up its excess energy as a flash of light. The colour of the light is characteristic of the gas. Neon gas gives the red light familiar in advertising signs. Sodium gas is very efficient at producing light but its strong yellow colour, though acceptable for street lights, is useless in the home. Remember it is *impossible* to produce white light by this means. However some gases give enough different colours to look whitish. Xenon is an example. Xenon also produces some invisible ultra-violet light. The tube has a fluorescent lining which absorbs the ultra-violet light, and emits visible light filling in some of the gaps in the series of colours given by the xenon. The final effect is very close to white light.

Left: Fluorescent lamps can be bent into a variety of shapes for convenience, or to serve the requirements of advertising.

Below: The light from a gas may have rather garish colours. The easiest way to achieve a pleasant whitish effect is to convert some of the colours, and all the ultraviolet, in a fluorescent lining on the inside of the tube.

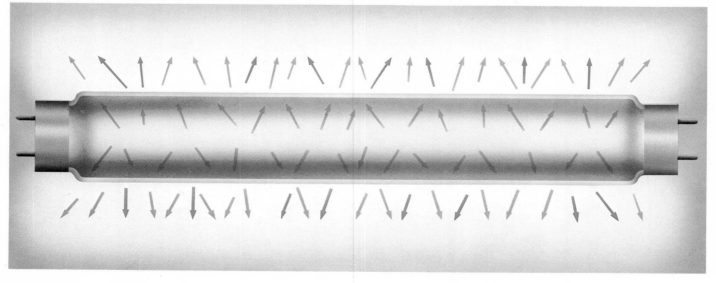

Supermagnets

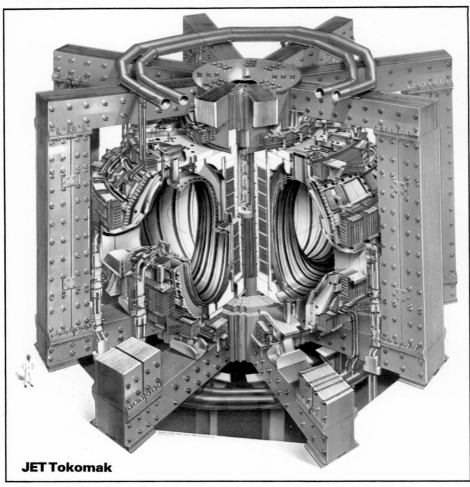

JET Tokomak

Above: By the end of this century, the world's biggest magnets will probably be those used to produce power by nuclear fusion. The Joint European Torus, a triumph of electromagnetic technology, is being built to contain a hot gas, or plasma, long enough and hot enough for the nuclei to react.

Below: The electron itself is a strong magnet. Atoms are magnetic if the electrons in them are lined up. That is the electrons spin in the same direction. Atoms lined up in groups can give rise to permanent magnetism, and this seems to work best in some complicated alloys and ceramics.

Helium atom

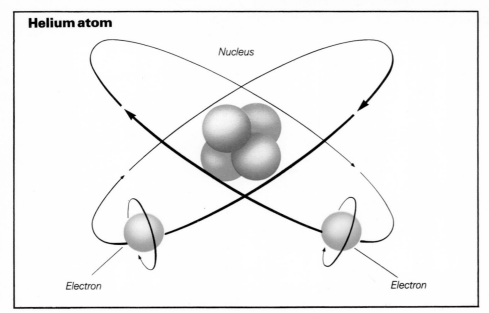

Nucleus

Electron

Electron

Coils versus permanent magnets

Which gives the strongest magnetic field, a very powerful permanent magnet or a very high current circulating through a coil?

A circulating current in a coil, or in the complicated convection of the Earth's interior, generates a magnetic field which threads through the loop and spreads out beyond it. The shape of the field produced by a coil is the same as that produced by a bar magnet.

In one sense there *is* a circulating current in a bar magnet. It is not made up of free electrons moving in a wire, but is the added effect of the electrons orbiting in the atoms. The orbiting electron represents a current, and produces a magnetic field threading through the atom. Electrons in atoms do not run down and lose their energy, as do electrons in a coil as soon as the battery is removed. This is why bar magnets can keep their magnetism permanently. Electrons in atoms only have a limited choice of orbits (see page 28), corresponding to the lowest energies and can stay in those orbits indefinitely.

In addition there is a second source of permanent magnetism. Electrons and protons are themselves magnetic, and so are the basic units of magnetism. Added together all these individual magnetic effects are very strong. If both types of magnetism for all the electrons in a bar magnet could be lined up with one another the magnet would be a thousand million times stronger than any yet made. But no-one has the slightest idea how to make the electrons line up in this way so the best answer to the question I posed at the beginning is to use a coil and a large current. In this way magnetic fields nearly a million times that of the Earth have been produced, but only for a brief time before the coil blows up. (A fully aligned permanent magnet would also explode; remember like poles repel each other and no known material would be strong enough to hold them together.)

Seen in this way the remarkable thing about iron magnets is not how they manage to be so strong but why they have not been made much stronger. One reason is that the atoms align themselves in small volumes called domains, but the domains are only weakly aligned with each other. The grouping is clearly seen on electron microscope pictures.

Below: Fermilab, west of Chicago in U.S.A., and CERN which straddles the French-Swiss border north of Geneva, are the sites of the world's biggest particle accelerators. The energy of the particles is far higher than in JET, so the magnets are in a much bigger circle, but the particles are confined to a tube only a few centimetres in cross-section inside the magnet. In both cases there are in fact several accelerators linked to each other. At Fermilab particles travel round a single ring, where as at CERN particles travel in opposite directions in two intersecting rings. The CERN particles are involved in violent head-on collisions at the intersections.

Fermilab

CERN

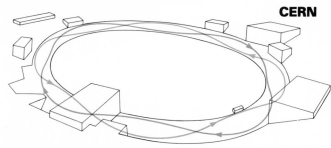

Magnets can be made from materials other than iron. The iron atom in itself is not especially magnetic; copper and aluminium atoms are almost as magnetic for instance. There have been a series of discoveries of alloys with ever better magnetic properties, and magnets are now adequate for most purposes. Who needs a more powerful loudspeaker or a smaller earphone? When digital meters take over, magnets for ammeters will not be needed at all. On the other hand there will always be a market for better magnets in small motors and generators. These at least seem unlikely to be superseded at the present time.

Electromagnets

When a varying magnetic field is required a permanent magnet is of little use. An electromagnet (coils wound on iron) must take its place. Atomic particle accelerators have given rise to some quite splendid electromagnets. The longest are for the machines at CERN in Geneva and Fermilab at Batavia in Illinois, which are both circular, 3km in circumference and remarkably slender in cross-section. The magnetic field varies over a factor of about 100 with a period of a few seconds.

Among the biggest magnets of all are those needed for machines producing nuclear fusion reactions. One of the most promising designs is the Tokamak invented in Russia. The largest prototype, the Joint European Torus, is being built near Oxford in England. It provides very nice examples of many of the principles of electricity, so is assured of a place in books even if it is not successful. The basic problem is that fusion reactions can only provide energy if the temperature is about 100,000,000 degrees. The idea is to contain this very hot gas by magnetic forces. The hot gas is called a plasma and it is prevented from touching the sides of the Torus by a combination of two magnetic fields at right angles to each other.

Storing electricity

Why store electricity?
There are two reasons for wanting to store electricity, *either* to run a portable appliance such as a radio or torch, *or* to store power when there is excess available and feed it back days, hours, or perhaps just seconds later when there is a peak demand. The only way of storing electrons themselves is in an electrical capacitor. A capacitor, at its simplest, is no more than a sheet of metal in which electrons can be held by an electric field. Capacitors are extremely useful in handling small rapidly varying current in electronic circuits, but are of no use on a large scale because only tiny amounts of energy can be stored. To store electricity we must therefore look for ways to store energy which can be converted efficiently into electricity.

Mechanical storage
Large amounts of energy can be stored in a mechanical form by pumping water up to a high reservoir. The water can be run back through a turbine to generate electricity when the main station cannot cope with demand.

A flywheel provides mechanical storage but on a much shorter time-scale. A flywheel is a large heavy wheel on the shaft of the generator which rotates with the shaft. It contains a lot of energy because of its rotation, and the stored energy keeps the shaft turning when there is a sudden increase in demand. Without the flywheel there would be a greater drop in speed, and therefore in voltage supplied to all the other users of the generator. A flywheel is only a partial cure for such unexpected peaks in demand as the occasional flickering of lights in the home shows.

Electrochemical storage
Energy is stored electrochemically in batteries. Some batteries are rechargeable like the car battery. Other batteries cannot be recharged very easily and are

Right: There are many different types of battery. The silver oxide battery can be made very small (a few millimetres across). It is also long lasting, so it is used in watches and cameras. The zinc carbon battery used in radios, calculators, torches and toys is usually larger (about 4 cm long and 2 cm across). It is usually discarded when discharged, although a limited amount of recharging is possible. A lead acid battery can be recharged many times, but is heavy and the acid is dangerous if it leaks or is spilt.

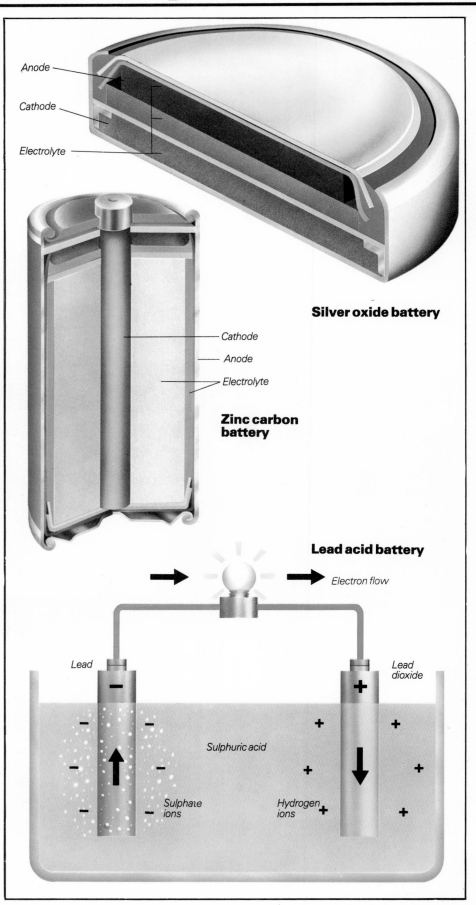

Silver oxide battery

Anode
Cathode
Electrolyte

Cathode
Anode
Electrolyte

Zinc carbon battery

Lead acid battery

Electron flow

Lead

Lead dioxide

Sulphuric acid

Sulphate ions

Hydrogen ions

thrown away after use like torch batteries. Whatever the type of battery they all depend on the fact that many chemical reactions involve the transfer of electrons from one atom to another atom. An atom or group of atoms which carries an electrical charge is called an ion. Batteries are built in such a way that positive ions collect in one part and negative ions in another part. These collection points are the terminals of the battery. If the terminals of the battery are connected together, through a light bulb for instance, a stream of electrons flows from the negative terminal to the positive terminal and the bulb lights up. The bulb will stay alight until all the electrochemical energy stored in the chemicals inside the battery is exhausted.

If an electric current is forced to flow through the battery in the opposite direction, the chemicals return to their original state. This is what happens when a car battery is recharged: energy is taken from the mains through a rectifier and stored inside the battery. Different types of battery have different chemicals inside them but the basic structure is the same: two collecting points called electrodes and a path between them provided by an electrolyte. The flow of ions happens because of reactions between the electrodes and the electrolyte. The car battery uses a series of lead oxide electrodes and sulphuric acid electrolyte and consequently is very heavy, and dangerous if knocked over, but can be easily recharged many times. Batteries used in torches have carbon as the positive electrode and zinc as the negative electrode. The electrolyte is a paste and so cannot spill. The batteries are very light and safe to use but inefficient to recharge and therefore expensive to use.

Electric cars

It would be very nice to be able to run electric cars from batteries which could be recharged overnight. They need to be lighter than the lead-acid-type because most of the energy would be used in moving the batteries around. Many novel systems have been tried but all have disadvantages. Increasing petrol prices and technological innovation may one day make the idea more attractive. Small vans for in-city use have had some success. They have the advantage of being very quiet and having no exhaust gases.

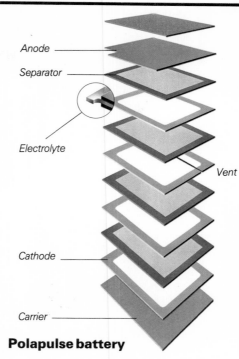

Anode
Separator
Electrolyte
Vent
Cathode
Carrier

Polapulse battery

Left: Cameras which produce on-the-spot prints employ a special film which has a battery incorporated into its structure. This battery is available as a separate product and can be used in toys, games, and any application which requires a thin power source.

Below: If cars ran off batteries, the roads would be quieter and less polluted. A number of experimental battery-driven vehicles have been made but so far the batteries are still too heavy and require recharging too frequently for a practicable motor car. The electric van in the photograph is used for around town light transportation. The diagram below shows the number and positioning of the batteries needed to power the van.

Electric van

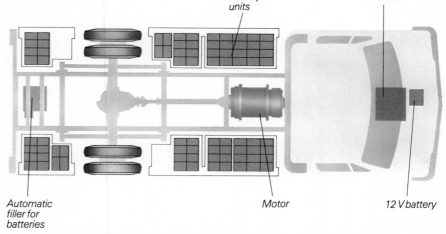

Battery units
Controller
Automatic filler for batteries
Motor
12 V battery

Nerve pulses

Nerves as cells

When Samuel Morse sat at his desk to write down the Morse code, the information to his fingers on how to move was also being carried along fibres in coded form; the pulses carried by nerves. By coding into a series of pulses, information or instructions can be carried in the most reliable way. Speech and music can be converted into pulses. The Post Office, and broadcasting and recording companies are all steadily moving towards pulse coding. This involves the replacement of existing copper cable with optical fibre cables which are much cheaper to manufacture than copper cables. Each fibre can carry thousands of telephone calls simultaneously in the form of pulsed light.

Electrical pulses in nerves are generated by chemical reactions with many of the characteristics of those in storage cells. Imagine a lead acid cell made in the form of a long tube with a small plate at each end. By adding a little lead oxide at one end, a voltage would be produced which could be detected at the far end. Indeed the corresponding

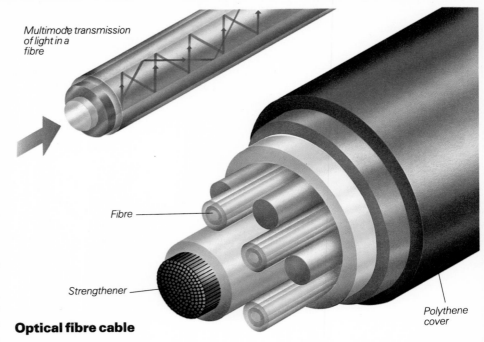

Multimode transmission of light in a fibre

Fibre

Strengthener

Polythene cover

Optical fibre cable

Below: Even the fastest athlete cannot react faster than the nerves carrying messages to and from the brain. That takes about a tenth of a second, during which time a tennis ball may travel 20 or 30 metres.

Above: Optical fibre cable contains many fibres and is capable of transmitting more messages simultaneously and more efficiently than copper cable. Pulsed light is immune from outside electrical interference

electrical charge could be made to release some lead oxide in an identical adjoining cell – and so on. Nerve cells work in much the same way, but use sodium and potassium salts instead of lead and acid. A set of several reactions which take place across the membrane of a nerve cell are so delicately balanced that they are disturbed by the slightest touch. The disturbance results (it is not known how) in the release of a tiny quantity of sodium, which gives a voltage of about 0.1 volt. (No wonder an electric shock of hundreds of volts is so painful and disruptive.) The pulse travels along the cell, and from one cell to the next, at about 100 metres per second. Thus a touch on your finger is registered in the brain a few hundredths of a second later. The brain can decide if action is necessary and send messages back to the hand in another few hundredths of a second. The muscles can start moving, thanks to further chemical reactions, in less than a fifth of a second. This reaction time is the limiting factor, not only in shoot-outs in Western movies, but in many sports such as tennis, cricket and boxing where success depends on the fastest possible response to your opponent's moves. Win or lose in a fraction of a second.

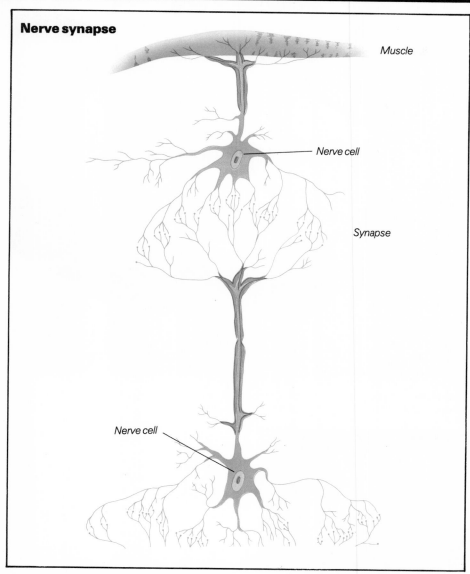

Nerve synapse

Muscle

Nerve cell

Synapse

Nerve cell

in ways that may perhaps never be understood in detail. One way that the brain is studied is by putting electrodes on the head (or even in the brain) to pick up voltages, but that is rather like trying to work out how a telephone exchange works by putting your ear to an outside wall. Nevertheless there are some overall characteristic rhythms of voltage which can be identified even in this simple way because they change in conjunction with different activities. These rhythms can be used diagnostically but only in the crudest ways. Their absence is used as a definition of death for instance. It is also possible to stimulate nerves directly to provide a *very* limited form of sight to a person with damaged eyes.

Electricity and growth

All growing cells show a small voltage difference across their outer membrane. The way animal life processes produce and use this voltage is not well enough understood to know how electricity should be used to stimulate healing or inhibit a cancer. But simple experiments have shown that mild currents can affect the healing of flesh wounds and speed up the mending of broken bones, so this might become a useful technique.

Below: The electrical activity of the brain is enormously complicated, but some simple characteristic rhythms predominate during certain states of rest or activity.

Above: A 'junction box' in a nerve is called a synapse. The electrical pulses from one nerve are passed on to the next nerve across the minute gaps.

Bundles of nerves

In the retina of the eye there is a mosaic of tiny special cells containing a substance which releases a charge when light falls on it. Each cell is connected to a nerve. The eye and the brain are linked by a bundle of thousands of these nerves, each a few thousandths of a millimetre thick and insulated from each other by very thin layers of fat called myelin. Somehow that vast mass of signals from the eye is sorted out by the brain into a picture. That is just one of the many tasks which the brain can carry out simultaneously. It contains some 10 billion cells, working together

Brain waves

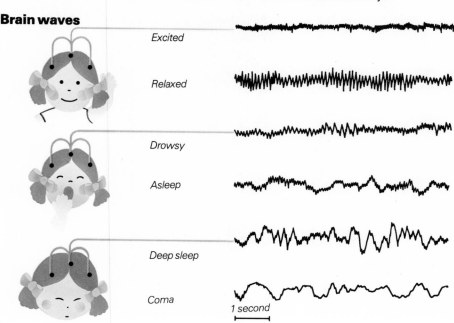

Excited

Relaxed

Drowsy

Asleep

Deep sleep

Coma

1 second

Plants and animals

Above: Migrating birds often travel at night and so cannot use the landscape for navigation. The Earth's magnetic field and, on clear nights, the pattern of the stars probably help them find their way.

Below: Plant roots contain a central woody section (xylem) in which the sap moves, and a surrounding of softer tissue. Nutrients cross the outer layer of the root selected in unknown ways by tiny electrical forces.

Plant root cross-section

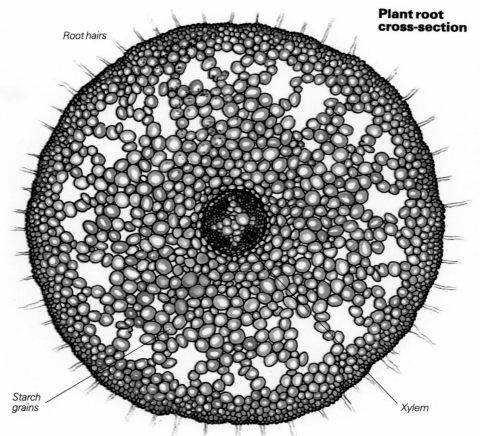

Root hairs

Starch grains

Xylem

Animal electricity

Man is often said to be unique among the animals. But human beings do many things badly in comparison with other animals: many animals of similar size or smaller run faster; the senses of smell and taste are blunted and hearing mediocre in man. Dolphins have forms of 'speech' which seem as sophisticated as ours, though we are not clever enough to tell what they are saying. Sharks and rays have an additional and separate means of communication by means of electric signals which other fishes can pick up through the water. (Try to imagine carrying on two conversations at once, one on the sound channel and one on the pulsed electrical channel.) They can detect tiny changes in electric field caused, for instance, by an intruder.

Some fish are capable of producing large pulses of electricity. The electric eel produces pulses of 500 volts and 1 amp, and does so under water which is a poor insulator. That is ample to blow up a large electric lamp, kill a horse and stun a man. It is lethal to other fish, but not of course, the eel itself. In the electric eel a large number of segments of modified muscle tissue (which both respond to and produce electrical signals), each providing only a few tenths of a volt, are timed by a network of nerves to add together to produce the staggering 500 volts. (The electric plates of the catfish are modified skin but act in a similar way.)

Navigation

Man's own senses give very little power of navigation. Electronic inventions have allowed an enormous improvement, but a few hundred grammes of swallow can do almost as well as several hundred kilograms of missile. Many birds have a remarkable ability (matched by some fish and turtles) to find their way, either over set paths as in annual migrations, or from any point back home, as in homing pigeons. Their methods are not well understood. They probably use the stars by night and recognise the landscape by day (as do Cruise missiles), but also sense the magnetic field of the Earth. Pigeons, and other animals, have magnetic particles in the ear, and are sometimes confused by magnetic storms, and by being transported to a region which is diametrically opposite to home in terms of the *magnetic* map of the Earth.

Plant electricity

In common with all animals, we depend directly or indirectly on plants for our food. Plants grow by taking up a carefully selected mixture of chemical nutrients into their roots. The force drawing in the nutrients is electrical. Plant root cells have a voltage of about 0.1 volts across them (negative inside the root) and even this small voltage

represents a very strong field across the very thin cell membrane. The field is well able to pull in *positive* ions (atoms with missing electrons) into the root. In this way potassium and other metals will be introduced into the plant. However the plant needs negative ions as well (chloride, nitrate and sulphate, to name only the simplest), and these have to overcome the electrical force. How this is achieved is not known. Possibly each negative ion is disguised in a large molecule and smuggled through a small pore in the membrane.

A few plants respond to touch in a way which looks at first sight to resemble an animal's reaction. The leaves of *Mimosa pudica* curl up in a few seconds when touched. But although this is an electrical response, plants certainly do not have a central nervous system like animals. The reaction is

Above: The coiling of the tendrils of climbing plants is thought to be an electrical response. This is the tendril of the grapevine, *Vitis vinifera*.

local, and not governed by a brain which receives the messages and decides whether and how to react. Your cat may either snuggle up to you or scratch you when you touch it, but your Mimosa will never do more than curl up its leaves. A less dramatic reaction to touch is very common in climbing plants which grow around objects which they touch, whether a thoughtfully placed bean pole, or a hedge within tendril distance of a convolvulus.

Below: The electric eel, *Electrophorus electricus*, can produce a large pulse of electricity for defence and immobilizing its food. Special battery cells sited along most of its body can all discharge together.

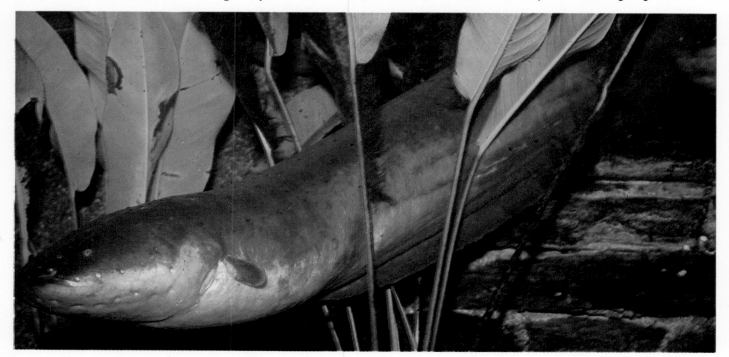

Fields in the sky

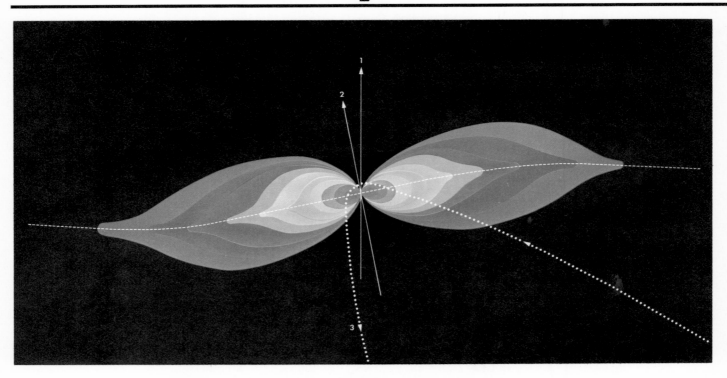

Below: A comet's tail is so flimsy it can be pushed aside by the solar wind. Comet Kohoutek (photographed from Earth in January 1974) showed a small kink in its tail, thought to be caused by a sudden change in the solar wind.

Below: This immense eruption in the sun's corona was photographed by Skylab's ultraviolet camera. The light comes from very hot helium streaming along the lines of magnetic force at the sun's surface.

Above: The effect of Jupiter's magnetic field spreads a long way into the surrounding space and deflects the solar wind. The diagram shows the rotational axis (1), the magnetic axis (2), and the path (3) of the space probe Pioneer 10.

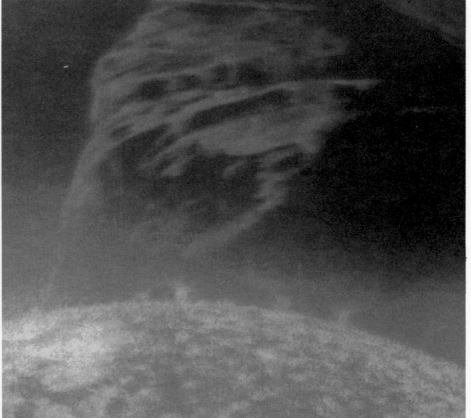

Splitting of light

Light source

Light source in a magnetic field

Light splits into two distinct bands

Influenced by magnetic field light splits into many more distinct bands

Prism

Current or rotation?

The Earth, as we well know, has a magnetic field. The Sun, as was guessed from the shape of the corona, has a magnetic field. Forty years ago it was only possible to test a handful of other stars for magnetic field, and they indeed showed quite strong effects. An eminent physicist, Professor P.M.S. Blackett speculated that *all* rotating bodies had a magnetic field, purely due to the rotation. The bigger the star, and the faster its rotation, the bigger the field. He set up sensitive magnetometers to check the much smaller fields that would, according to his idea, be generated by rotating bodies in a laboratory. The magnetometers never moved off the zero reading.

Nowadays many stars have measured magnetic fields, many more show very strong radio radiation, thought to arise in a magnetosphere outside the star. But the rotation idea is dead, and all materials lose *permanent* magnetism above a few hundred degrees centigrade. So as far as we know all these magnetic fields must be generated by electric currents inside the star.

Types of current

If you have read this book *very* carefully you will have sorted out four ways that currents can be produced. (If you want a hint they are on pages 4, 26, 28, 32.)
1. Electrons force their way through air as a result of a high electric field. Their collisions dislodge more electrons which carry a current and cause a spark.
2. Electrons are freed from atoms, in the particular conditions in metals, and so carry a current.
3. Electrons are freed from atoms, due to collisions in a gas at high temperature, and so can carry a current. A hot gas in this form is called a plasma.
4. Electrons and ions travel in opposite directions, carrying half the current each, as in a storage battery.

The Earth's magnetic field is due to a process similar to (2), with the rocks in the core taking a metal-like structure because of very high pressure. The Sun's field certainly arises from plasma currents, as in (3) surging through the Sun and into its atmosphere. No doubt similar processes take place in many other stars. The fields of very dense

Above: A magnetic field changes the pattern of wavelengths of light from a source. The magnetic fields of stars can be measured thanks to this effect.

collapsed stars called white dwarfs with a mass like the Sun but a radius like the Earth (100 times less than the Sun) are more difficult to explain. The material is nothing like rock and has a density a million times that of the original star, and the magnetic field is increased by a corresponding amount.

Other natural currents

Does nature use the other two ways of making electric current? There is an example of the process (1) on Venus, which has thunderstorms, probably much like our own. They have been recorded by an artificial satellite orbiting Venus.

The type of electrical reactions taking place in plant and animal cells and man made storage cells may just possibly be unique on Earth. They need *liquid* water (or at least a jelly) and that is one of the rarest substances in the Universe.

Reversible electricity

Making plastic bags

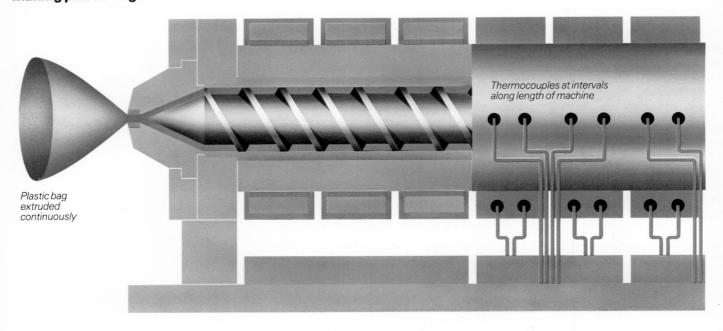

Thermocouples at intervals along length of machine

Plastic bag extruded continuously

Above: Thermocouples measure differences of temperature, and can penetrate inaccessible places. This is just what is needed to monitor the cooling of plastic in an injection moulding machine.

Left: The solar panels of Skylab converted sunlight to electricity to power the on-board electrical equipment. One of the panels was damaged during take-off and this accounts for the unbalanced aspect.

Efficiency

Imagine a pumped storage system (see page 19), viewed from so far away that none of the working parts can be seen. Sometimes water would flow out and electricity be produced, sometimes the reverse would be the case, with electricity being used to pump water in. A very careful observer might notice that when water flowed out it never produced quite as much electricity as was needed to pump it back. The ratio of the electricity produced to the electricity used up is called the efficiency. There are a number of such almost reversible transfers which take place inside matter, or even inside atoms. We cannot see what is happening directly, but physicists can work out the processes.

Thermocouple

A current flows in a thermocouple circuit when the two junctions are at different temperatures. To reverse this system, pass current through the thermocouple junctions. One will be heated, which may not seem surprising, but the other will be *cooled* by the passage of the current. This is not just a scientific curiosity because it is put to use in cooling small components in electronic circuits. A near perfect refrigerator with no moving parts.

Solar cell
When light falls on a solar cell it produces a voltage and, connected into a circuit, can power a photographer's light meter or a solar powered watch. Many such cells can power a satellite. The reverse process, in which passing a current through the material produces light without heat can sometimes occur and is called electroluminescence.

Quartz watch
The pumped storage system is a transfer between mechanical and electrical energy and there is an atomic example of that type of transfer. The atoms in a crystal are held in place, in such positions that the electrical forces between one atom and all its neighbours are in balance. If the crystal is deliberately distorted from its natural shape by squeezing it, the electrical forces are no longer in balance and a voltage will develop across the crystal. Quartz is a particularly good material to use. The effect is called the piezoelectric effect and is put to use in record player pick-ups. The stylus vibrates a crystal when it passes along the groove in the record. The voltage which the crystal develops is amplified and converted to sound.

The reverse of this effect is that an alternating voltage will compress or expand a crystal, depending on whether the voltage is positive or negative. In a quartz watch a crystal vibrates, and produces an alternating voltage which after amplification runs the motor that drives the watch. But the crystal itself must be *kept* vibrating, so part of the amplified signal is fed back to the crystal, keeping it topped up with mechanical energy. Thus the energy transfers in both directions are essential to keeping a quartz watch going.

Right: The *Solar Challenger* crossed the English Channel in 1981 powered by 16,000 solar cells. It is a dramatic demonstration of solar energy, but there is no prospect of making a commercial airliner powered in this way

Watches commonly use a crystal which vibrates 32,768 times per second, but better ones use 2,097,152 Hz because the note is even purer. If these numbers seem strange, just keep dividing by two, and remember that is what the watch has to do to end up advancing the hand (or figures) once per second.

Filters
The smaller the crystal the higher the frequency at which it vibrates. Quartz crystals can be used to accurately pick out a desired frequency in a radio set. This use is totally different from the use of the crystal in that pioneer of radio the *crystal set*. There the crystal rectifies the radio waves. The filtering effect is used at sound frequencies in submarine telephone cables, by setting up a wave which travels over the surface of the crystal rather than vibrating the whole crystal.

The future

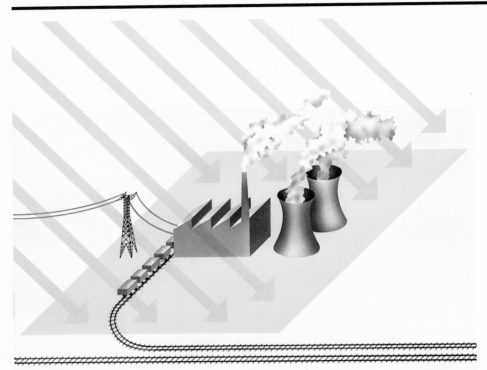

Left: The power of solar radiation falling on a power station site, on a sunny day, is as much as the power generated by that station. Unfortunately solar energy conversion cells are still much too expensive for commercial power production.

Use what we have

There are already so many clever, useful, efficient or just plain enjoyable ways of using electricity that the main advance in the future ought to be rather ordinary: to spread the existing appliances to more of the world's peoples. Even in the most industrialised countries there are a large number of people who do not exploit the full advantages of electrical devices. Also we could all benefit from better and cheaper lighting, better cookers, more power tools, and so on.

Although oil reserves are limited, and coal mining is dangerous there is no *fundamental* difficulty in obtaining power from other sources. Tropical countries have abundant solar energy which can be harnessed by fermenting or burning plants if the rainfall is high, or by direct conversion in photovoltaic cells if the rainfall is low. In temperate countries, if the rainfall and mountains are high, there is hydroelectric power. Flat dry countries may be windy and coastal countries can exploit tidal and wave energy. Nuclear fuel does not quite count as inexhaustible though the fusion reactor would use commonly available materials, if it ever becomes

Right: Plants are very efficient in using solar energy to manufacture plant tissue. Some of that energy can be released in a usable form by fermenting the tissue to produce alcohol which can be used in place of petrol.

technically possible (see page 31).

There is an enormous scope for more efficient use of the power which is produced. Thermal power stations, which are unfortunately going to continue to burn up valuable resources for a long time yet, waste at least half the power in the form of heat. Whilst it is impossible to avoid the heat as a by product, there are plenty of ways it can be used by piping it to local homes and industries. This is very rarely done but has been very successful in cities, such as Rotterdam, that have taken advantage of the possibility.

Distribution

There is another major power loss in the *distribution* of electricity: roughly three per cent per hundred kilometres. If our imaginary country with 50 power stations (see page 18) is 1,000 kilometres square, the stations will be 140 kilometres apart, if they are evenly spaced. In practice the sources of power have an annoying habit of being sited even further from heavily populated areas. There is little scope for reducing the losses through heating in conventional cables. Such losses could be eliminated by cooling cables to within a few degrees of the absolute zero of temperature (−273°C) in which case electrical resistance disappears altogether. The engineering complexity of the cooling system unfortunately means that such *superconducting* cables could only be financially worthwhile for short heavily used sections. They are unlikely to be economic over large areas of country.

Magnetic storage

Superconductivity can be used in a quite different way. Storage of energy improves efficiency, but so far only pumped water storage is financially worthwhile. Using superconducting coils it is possible in principle to store a vast amount of magnetic energy. It needs a huge magnet, concreted into an underground cave to withstand the high magnetic forces which tend to push the coils apart. It sounds rather fanciful, but remember that the *newer* sources of energy from wind, waves

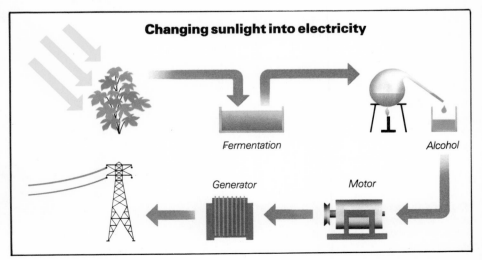

Changing sunlight into electricity

Fermentation   *Alcohol*

Generator   *Motor*

Energy used per person

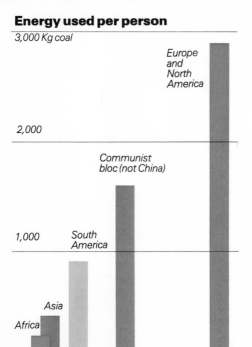

3,000 Kg coal

Europe
and
North
America

2,000

Communist
bloc (not China)

1,000

South
America

Asia

Africa

0   Income/person $ 1,100

Above: The countries which use more power per person also pay higher wages per person. Which is the cause and which is the effect?

Right: The considerable losses of energy in power transmission cables could be eliminated by using superconducting cables. Such cables operate at a very low temperature and have no resistance at all, but unfortunately are very expensive to make and to keep cold.

and sunlight are very variable, so storage of energy is going to be an increasing problem.

New applications

The really useful projects in the future are likely to be concerned with linking biology and electricity. Here are a few examples which have been mentioned in this book. I hope I have given you enough curiosity and interest to find out more about them.

1. Generation of electricity through biological means, directly and through fermentation.
2. Medical use of electricity.
3. Electricity as a stimulant to plant growth.
4. Direct connection to human nervous system.

Right: Half the energy output of our power stations is wasted as steam. If that steam were used to heat nearby buildings, electricity could be *far* cheaper.

Superconducting cable

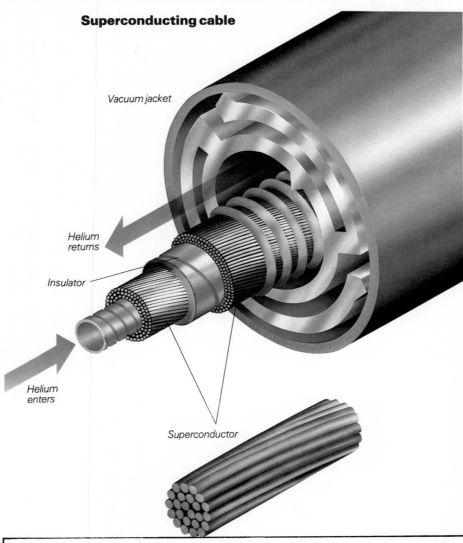

Vacuum jacket

Helium
returns

Insulator

Helium
enters

Superconductor

Using waste heat

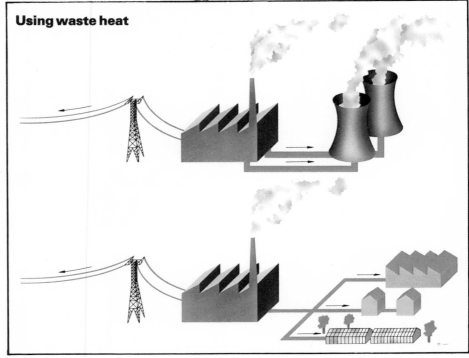

A-Z Glossary

Alloy is a mixture of metals, produced by melting them together.

Alpha rhythm is a regular series of electrical pulses in the brain.

Alternating current (a.c.) is a current varying smoothly and reversing its direction of flow with a given frequency.

Ammeter is any instrument which measures current, but usualy consisting of a coil in a magnetic field.

Anode is an electrode which is held at a positive voltage and therefore attracts negative particles.

Atom is the smallest unit of an element and consists of negatively charged electrons and a positively charged nucleus.

Auroras are veils of light produced high in the atmosphere in polar latitudes caused by bombardment of particles from the Sun.

Bacteria are single-celled organisms.

Breaker is a spring-loaded switch which is tripped by an excess current and breaks the circuit, but can be reset.

Capacitor is a pair of electrodes separated by a small distance, capable of storing small amounts held in place by electrical forces.

Cell consists of two plates normally of different metallic compounds, immersed in a solution of a salt, producing an electrochemical voltage.

Circuit is a pathway for free electrons through or along conductors.

Conduit is a tube containing, and protecting, electrical cables.

Corona is the outer atmosphere of the Sun. It is very tenuous, but very hot, and extends several hundred thousand miles.

Crystal is a solid with a characteristic shape arising from the regular array of its molecules.

Current is electrical charge flowing down a wire or through space.

Cut out is a breaker incorporated in a multimeter.

Diaphragm is a thin tightly stretched sheet.

Direct current (d.c.) is current flowing continuously in the same direction.

Domain is a small region in a ferromagnetic material in which the molecules are magnetically aligned.

Earth leakage trip is a breaker in the wire of a house system which is operated by excessive current.

Electrodes are the plates in batteries or capacitors.

Electromagnetic waves are waves such as radio and light waves which travel freely through space.

Electron is a tiny particle with a negative charge.

Electronics is the science of electrons in controlled motion.

Filament is a thin wire in an incandescent bulb.

Fluorescence is the process by which some materials can absorb light of one colour and later emit light of another colour of longer wave length.

Fluorescent tube is a tube in which a gas at low pressure transmits a current with the emission of light, some of which is converted to another colour by a fluorescent coating on the inside of the tube.

Fuse is a thin wire which melts (fuses) and so breaks if the current exceeds a given limit.

Grid is the network of cables distributing power across the country.

Hertz (Hz) is the name given to the unit 'one cycle per second'. Mains power is transmitted at 50 Hz or 60 Hz depending on the system used by the country concerned.

Hydroelectric power is electrical power derived from the mechanical power of flowing water.

Incandescent lamp is a lamp in which the light is provided by a thin wire heated by electricity. The wire is called the filament.

Insulation is a covering, made of material of very high resistance, for a wire or cable, or even a nerve cell.

Integrated circuit is a large number of transistors and other electronic elements in a very compact arrangement.

Ion is an atom which has been charged positively by the removal of an electron, or negatively by the addition of an electron.

Lead acid accumulator is a system of lead plates and sulphuric acid to store electricity, widely used in cars.

Leclanché cell is a system of zinc, ammonium chloride and carbon commonly used in torch batteries.

Lodestone is a naturally occurring magnetic material, magnetite.

Magnet is a bar or horseshoe of iron alloy which retains strong magnetic properties, but used more generally for any magnetic body.

Magnetic field is the pattern of magnetic force near a magnet.

Magnetic pole is one end of a magnet.

Magnetic storm is rapid changes in the Earth's magnetic field caused by streams of charged particles (electrons and ions) from the sun.

Magneto is a generator used to provide electricity for the spark plugs in a small motor.

Membrane is any thin layer of separation, but used particularly for outer layer of a cell.

Microphone is a device to change sound waves into (alternating) electrical current.

Microwave is a radio wave with a frequency in the region of a thousand million hertz.

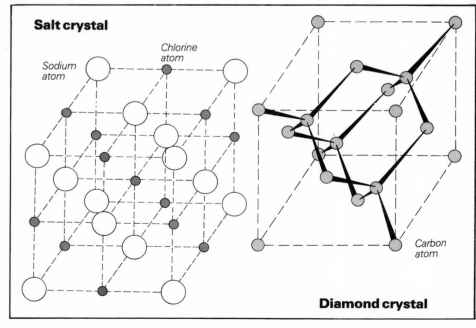

Salt crystal

Sodium atom

Chlorine atom

Carbon atom

Diamond crystal

Inside a colour television tube

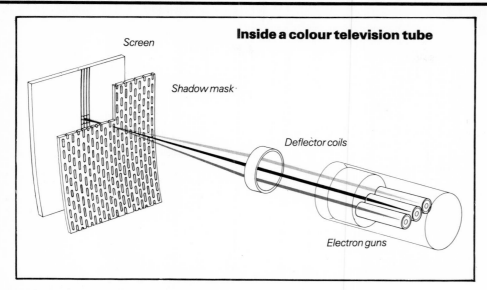

Screen

Shadow mask

Deflector coils

Electron guns

Migration is the seasonal movement of birds (or sometimes animals) often over thousands of miles.

Morse code is a system in which each letter is assigned an agreed pattern of dots and dashes.

```
A  · —          N  — ·
B  — · · ·      O  — — —
C  — · — ·      P  · — — ·
D  — · ·        Q  — — · —
E  ·            R  · — ·
F  · · — ·      S  · · ·
G  — — ·        T  —
H  · · · ·      U  · · —
I  · ·          V  · · · —
J  · — — —      W  · — —
K  — · —        X  — · · —
L  · — · ·      Y  — · — —
M  — —          Z  — — · ·

1  · — — — —    6  — · · · ·
2  · · — — —    7  — — · · ·
3  · · · — —    8  — — — · ·
4  · · · · —    9  — — — — ·
5  · · · · ·    0  — — — — —
```

Multimeter is an ammeter, with built-in resistors and cells, which can measure current, voltage and resistance.

Myelin is the coating which insulates nerve cells.

Neon is a gas which emits red light when a current is passed through it.

Neutron is a particle slightly heavier than the proton but with no electrical charge.

Nucleus is the central part of an atom, made up of protons and neutrons.

Nutrient is the solution of chemicals needed to feed a plant.

Ohm's Law is the proportionality of current and voltage which applies to many, but not all, materials.

Orbit is the path of one body around another when there is a force attracting the two together.

Pick-up is a small head which senses the shape of the groove in a gramophone record.

Piezoelectric voltage is the voltage produced when special types of crystals are squeezed.

Power is the rate at which energy is being supplied, equal to current multiplied by voltage, also for most devices equal to voltage multiplied by itself and divided by the resistance.

Power station is a place where a considerable amount of power (from any source) is converted to electricity.

Proton is a tiny particle with positive charge.

Rectifier is a device which allows current to pass in only one direction.

Resistance is hindrance to the passage of a current. Small resistors are used to do this in an accurate and controlled way.

Semiconductor is a material with a resistance higher than that of metals, but not as high a resistance as an insulator. Such materials have additional properties making them useful in transistors.

Spin is the circular motion of a body about its own axis and, by extension of meaning, the energy of circular motion.

Storage reservoir is a small high reservoir which stores excess power station output in times of low demand and releases it in times of peak demand.

Sulphuric acid is a powerful acid used in batteries.

Superconductivity is the disappearance of resistance in some materials at very low temperatures.

Telegraphy is communication by coded electrical signals along wires.

Television tube is a vacuum tube which has a coating on the inside face. This coating is stimulated to give off light by a beam of electrons scanned rapidly across and down the face.

Telex is telegraphy in which the message is automatically coded and typed by modified typewriters.

Thermocouple is a pair of wires of different metals which produce a voltage proportional to the temperature difference between its ends.

Transformer is a pair of coils on an iron core used to change the voltage of an alternating current.

Transistor is a tiny device made up of slices of different semiconductor, and capable of controlling and therefore amplifying electrical signals.

Units of measurement such as metre, kilogramme, volt, ampère and ohm are subdivided by an agreed series of prefixes: milli = 1/1000, kilo = 1000, mega = 1,000,000, giga = 1,000,000,000.

Van de Graaff generator is a machine with a continuous belt taking charge up to a spherical storage terminal and producing very high voltages.

VHF is a band of radio waves with frequencies in the region of a hundred million hertz.

Voltage is the unit in electricity corresponding to pressure in water.

Voltmeter is a meter which indicates voltage across an element in a circuit.

Watt meter is a meter which records the power in use (a rather rare type of meter).

Watt hour meter is a meter, used in every home, to record the total amount of energy used. (Energy = power × time)

Reference

People and events

The following are some of the important dates connected with man's understanding and use of electricity.

c.1100 Use of lodestone as a magnetic compass (there may be unrecorded use much earlier).

1600 Gilbert introduces the idea of electric force, and suggests that negative electrical charge can flow from one body to another.

1724 Stephen Gray distinguishes between conductors and insulators.

1746 The idea of the earth return recognised by Winckler.

1752 Lightning recognised as a manifestation of electricity by Benjamin Franklin.

1780 Galvani identifies currents of electricity.

1790 Henry Cavendish, millionaire recluse

related to the Duke of Devonshire, discovers the laws now attributed to Coulomb and Ohm, and studies the use of the capacitor for storing charge.

1802 Coulomb published the law of attraction of positive and negative charges (and repulsion of like charges) with a force decreasing with distance between them (inverse square law).

1820 Magnetic effect of a current discovered by Biot and Oersted.

1825 Sturgeon invents the electromagnet.

1827 Ohm publishes the proportionality of voltage and current discovered by Cavendish.

1831 Voltage produced by varying magnetic field discovered by Faraday, and used in the first transformer.

1833 Faraday shows that all forms of electricity are the same.

1835 Electric telegraph, and code for its use, invented by Morse.

c.1850 The speed of light measured by Fizeau and Foucault.

1860 Lead/acid cell (now used in car batteries) invented by French physicist Gaston Planté.

1866 Dry cell now used in torches and radios invented by French chemist George Leclanché.

1871 Z. Gramme, a Belgian, made the first commercial generator.

1873 Maxwell wrote a book combining all previous work on electricity and magnetism and predicting the existence of a wide spectrum of electromagnetic waves including light.

1876 P. Jablochoff, a Russian living in Paris, made the first carbon arc lamp and Alexander Graham Bell transmitted the first telephone message to his assistant in Boston, U.S.A.

1877 Edison invented the phonograph.

1878 Microphone demonstrated by D.E. Hughes at the Submarine Telegraph Company offices, London.

1879 Edison makes greatly improved types of incandescent lamp in the first industrial research laboratory which he built at Menlo Park, U.S.A. First detection of radio waves by Hughes.

1880 Pierre Curie discovers piezoelectricity.

1881 First public electricity supply, Godalming, England.

Right: Marconi is seen here in 1901 with receiving and transmitting apparatus similar to that used for the first radio signal across the Atlantic. The signal, the three dots for *S* in Morse, were transmitted from Cornwall, England, and picked up in St John's, Newfoundland, in December of that year.

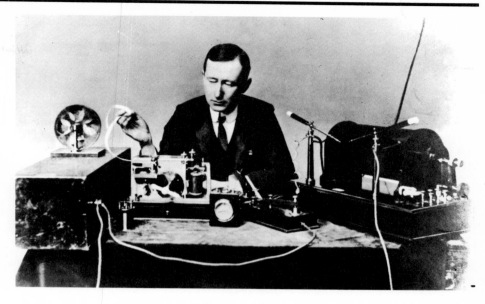

Left: Thomas A.W. Edison, seen here posing in his laboratory at Menlo Park, New Jersey, U.S.A., was undoubtedly the most prolific inventor of all time. He patented almost 1300 machines in his long and successful career and founded the Edison Electric Light Company in 1878 when he was thirty-one years old.

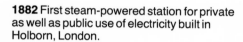

1882 First steam-powered station for private as well as public use of electricity built in Holborn, London.

c.1885 Radio wave properties described by Hertz.

1890 First electric locomotive in public service on underground railway in London.

1895 Marconi begins series of experiments on signalling by radio.

1897 Discovery of the electron by Thomson.

1901 Marconi sends first transatlantic radio message. Hubert Cecil Booth invents the vacuum cleaner in London.

1904 Photoelectric cell developed in Munich by Arthur Korn for scanning photographs enabling them to be transmitted by telegraph.

1908 Electric vacuum cleaner introduced in U.S.A.

1910 French physicist George Claude demonstrates neon light at the Paris Motor Show.

1911 Superconductivity discovered.

1935 Fluorescent lighting first demonstrated by General Electric Company in U.S.A.

1948 Invention of transistor by Bardeen, Shockley and Brattain, opening the way to all modern electronics.

1954 First nuclear power station established at Obninsk, USSR, and began generating electricity for industry and agriculture.

1956 First power for public use from a nuclear power station on a regular basis, from Calder Hall, England.

1957 Theoretical explanation of superconductivity.

1960 Explanation of electrical pulses in nerves by Hodgkin and Katz.

1977 First power fast breeder reactor on load at Dounreay, Scotland.

1981 American aircraft *Solar Challenger* crosses English Channel powered by 16,000 solar cells.

Books to read

How Did We Find Out About Electricity? by Isaac Asimov (Walker and Co., 1973)
Understanding Electricity and Electronics by Peter Buban and Marshall L. Schmitt (McGraw-Hill Book Co.)
Electric Power by Ed Catherall (Silver Burdett, 1981)
Electricity by Phil Chapman (EDC Publishing, 1976)
The First Book of Electricity by Sam and Beryl Epstein (Franklin Watts, 1977)
The Race for Electric Power by Jerry Grey (Westminster Press, 1972)
Using Electricity by Ronald J. Hamilton (Prentice Hall, 1971)
Power by Michal Kentzer (Silver Burdett, 1982)
Watt's Happening by Joni Keating (Good Apple, Inc., 1981)
Magnets, Bulbs, and Batteries (International Book Centre)
Wires and Watts: Understanding and Using Electricity by Irwin Math (Charles Scribner's Sons, 1981)
Men and Discoveries in Electricity by B. Morgan (Transatlantic Arts Inc.)
Rediscovering Electricity by Eugene F. Provenzo Jr. and Asterie B. Provenzo (Oak Tree Publications)
Electricity Experiments for Children by Gabriel Reuben (Dover Publications)
The Wonder of Electricity by Hyman Ruchlis (Harper and Row, 1965)
Fun With Electricity by Colin Siddons (Sportshelf and Soccer Associates)

Acknowledgements

Artist
Michael Robinson

The author and publisher wish to thank the following organisations for providing references for artwork on the pages indicated.

ADC, 15T
AKG Acoustics Ltd, 15BR
Chloride Technical Ltd, 33B
Mallory Batteries Ltd, 32
Open University, 43TL
Polaroid UK Ltd, 33T
Suunto Marine Compasses, 5
The Acoustical Manufacturing Co Ltd, 4B
The Electricity Council, 24-25

Photographs
Alan Cooper, 44, 45
Aldus Archive, 6
Avo Ltd, 13
Biofotos, 37T
Paul Brierley, 2-3, 9
CEGB, 18, 20
CERN, 31R
Fermilab, 31L
GEC Distribution Transformers Ltd, 17
GEC Turbine Generators Ltd, 14
JET Joint Undertaking, 30
Gary Ladd/Nicholas Hunter Publications, 4
Frank Lane Agency, 37B
Mansell Collection, 46, 47
NASA, 38, 40
Open University Archive, 7
Osram-GEC Ltd, 25B, 29
Renault Trucks/Karrier Motors Ltd, 33
Rex Features, 41
Science Picture Library/Dr Tony Brain, 25T
Tony Stone Associates, 36
ZEFA, cover, 8, 11, 12-13, 28, 34

Key: T (top); B (bottom); L (left); R (right)

Index

3 4 5 6 7 8 9 10—U—90 89 88 87 86